ISLANDS IN THE RAINFOREST

NEW FRONTIERS IN HISTORICAL ECOLOGY

Dynamic new research in the genuinely interdisciplinary field of historical ecology is flourishing in restoration and landscape ecology, geography, forestry and range management, park design, biology, cultural anthropology, and anthropological archaeology. Historical ecology corrects the flaws of previous ecosystems and disequilibrium paradigms by constructing transdisciplinary histories of landscapes and regions that recognize the significance of human activity and the power of all forms of knowledge. The preferred theoretical approach of younger scholars in many social and natural science disciplines, historical ecology is also being put into practice around the world by such organizations as the UNESCO. The series fosters the next generation of scholars offering a sophisticated grasp of human-environmental interrelationships. The series editors invite proposals for cutting edge books that break new ground in theory or in the practical application of the historical ecology paradigm to contemporary problems.

Series Titles

Vol. 1: *Social and Ecological History of the Pyrenees: State, Market, and Landscape*, Ismael Vaccaro and Oriol Beltran, eds.

Vol. 2: *The Ten-Thousand Year Fever: Rethinking Human and Wild Primate Malarias*, Loretta A. Cormier

Vol. 3: *Sacred Geographies of Ancient Amazonia: Historical Ecology of Social Complexity*, Denise P. Schaan

Vol. 4: *Islands in the Rainforest: Landscape Management in Pre-Columbian Amazonia*, Stéphen Rostain

ISLANDS IN THE RAINFOREST

Landscape Management in Pre-Columbian Amazonia

Stéphen Rostain

Translated by Michelle Eliott

Left Coast Press Inc.

Walnut Creek, California

LEFT COAST PRESS, INC.
1630 North Main Street, #400
Walnut Creek, CA 94596
http://www.LCoastPress.com

ISBN 978-1-59874-634-1 hardback
ISBN 978-1-59874-636-5 institutional electronic
ISBN 978-1-61132-687-1 consumer electronic

Library of Congress Cataloging-in-Publication Data:

Rostain, Stéphen,
Islands in the rainforest : landscape management in pre-Columbian Amazonia / Stéphen Rostain ; translation by Michelle Eliott.
p. cm.— (New frontiers in historical ecology Vol. 4)
Includes bibliographical references and index.
ISBN 978-1-59874-634-1 (hardback : alk. paper) — ISBN 978-1-59874-636-5 — ISBN 978-1-61132-687-1
1. Indians of South America—Agriculture—French Guiana. 2. Indians of South America—Land tenure—French Guiana. 3. Raised field agriculture—French Guiana—History. 4. Indigenous peoples—Ecology—French Guiana. I. Eliott, Michelle. II. Title.
F2460.2.A37R68 2012
980'.01—dc23
2012026551

Printed in the United States of America

⊗™ The paper used in this publication meets the minimum requirements of American National Standard for Information Sciences—Permanence of Paper for Printed Library Materials, ANSI/NISO Z39.48–1992.

Cover design by Piper Wallis
Cover photo by Stéphen Rostain

To Belém,
and to the memories of James Petersen, a wonderful companion, an admirable professor, and an ideal colleague; and Claude Muszynski-Delpuech, a partner in Caribbean archaeology who is dearly missed.

Contents

Illustrations

Acknowledgments

My research has always been based on strong collaborations and an interdisciplinary perspective, and so the success of this work is due to the contributions of a number of people and institutions over the last 10 years, to whom I am grateful:

- The participants of the Action Collective de Recherche (ACR) "Prehistory of the West Coast of Guiana" (2003–2007).
- The participants of the Earthmovers consortium, who provided a great deal of indispensible data during the "Archaeology and ecology of the coastal savannas of Guiana" project (2007–2011). I am particularly grateful to Laure Déodat of the Unité Mixte de Recherche (UMR) 8906 for her help in preparing the maps.
- Manuel Arroyo-Kalin, British Academy Postdoctoral Fellow at the University College London, for his photographs.
- Arie Boomert, emeritus professor at the University of Leiden, for providing photographs and documents.
- Frans Bubberman, forester in Suriname, for providing photographs and documents.
- Gérard Collomb, researcher at Laboratoire d'Anthropologie des Institutions et des Organisations Sociales (LAIOS), Centre National de la Recherche Scientifique (CNRS), Paris, for providing useful documents.
- Philippe Descola, director of the Laboratoire d'Anthropologie Sociale (LAS) and professor at the College of France (Paris), for his longstanding support.
- Émilie Dotte, doctor of archaeology and archaeobotanist at the University of Paris 1 and the Australian National University, for her photographs.
- Elisabeth Ewstifeieff for always providing a warm welcome at Montjoly, French Guiana.
- Serge Guiraud for his photographs.
- Michael Heckenberger, associate professor at the University of Florida, for his documents.
- Doyle McKey, professor at the University of Montpellier 2 and member of the Centre d'Ecologie Fonctionnelle & Evolutive (CEFE)

(UMR 5175), and a brilliant colleague, for initiating a wonderful project and for all of his advice. He is a *maître-d'œuvre*, admirable not only in the domain of interdisciplinary science, but also in the communal dynamics of teamwork. His comments on the fieldwork and this manuscript were indispensable.

- Jean-Pierre Montoroi, researcher at the Institut de Recherche et de Développement (IRD), for his photographs.
- Eduardo Góes Neves, professor at the University of São Paulo, Brazil, for his documents and his friendship.
- José Oliver, reader at the Institute of Archaeology of London, for sending photographs.
- Alain Pavé, professor at the University of Lyon 1, for his support of the "Archaeology and Ecology of the Coastal Savannas of Guiana" project.
- Martti Pärssinen, professor at the University of Helsinki, for his photographs.
- Francisco Valdez, researcher at the IRD, for his photographs.
- John H. Walker, assistant professor at the University of Central Florida, for his photographs.
- The Centre Spatial Guyanais, and particularly Sandrine Richard and Philippe Delaunay, for authorizing and facilitating the work carried out within the secure zone
- The CNRS for financing the "Archaeology and Ecology of the Coastal Savannas of Guiana" project from 2007 to 2011.
- The Ministère de la Culture et de la Communication, for financing the ACR project and the excavations carried out from 2003 to 2009.
- The Museum of Archaeology and Anthropology at Cambridge University, for opening its doors to me and hosting me.
- The Musée des Cultures Guyanaises, for allowing me access to its collections.
- The Fort Zeelandia Museum at Paramaribo (Suriname), particularly Laddy van Putten and Hanna van Petten, for their always warm welcome and access to the collections.
- The UMR 8096 "Archaeology of the Americas" (CNRS), where I carry out my research, and particularly Chantal Thomas, who finds solutions for all problems.
- Françoise Rostain, *qui m'apprit à me réveiller*.

Foreword

I first had the pleasure of meeting Stéphen Rostain 16 or 17 years ago, when I was a member of the scientific council of the French Institute of Andean Studies (Institut Français d'Études Andines [IFEA]), striving to promote the archaeology of upper Amazonia within this institution. Despite the impetus given by renowned researchers, particularly Betty Meggers and Donald Lathrap, few had taken up the call to enter this field of research, and it had fallen into neglect. However, it was obvious to me, as it was to most other ethnologists specializing in Amazonia throughout the world, that the new frontier of knowledge of the cultural specificities of this vast area, more comparable to a continent than a simple region, lay in a better knowledge of its precolonial past. In the words of Claude Lévi-Strauss in *The Raw and the Cooked,* Indian America was "a kind of Middle Ages which lacked a Rome." It was therefore necessary to try to uncover a little of this ancient history to shed more light on the particularities of contemporary Amazonian societies.

In a more selfish vein, my interest in the archaeology of upper Amazonia was also stimulated by the desire to advance the knowledge of the past of the region where I had conducted my own ethnological research, the basins of the northern tributaries of the Marañon, and especially the upper Pastaza valley occupied by the Achuar Jivaros. In most contemporary habitat sites, abundant potsherds and areas of anthropogenic black earth seemed to indicate that, despite very low current densities of human occupation, the inhabitants of the region had continually resettled the same sites over a long period. In addition, my work had shown that, despite the probably rather longstanding occupation of two distinct ecological niches, the poor soils of the *terra firme* and the rich alluvial terraces of the Pastaza, the Achuar had not, contrary to the predictions of Betty Meggers, developed clearly contrasting cultural styles or forms of use of the environment (Descola 1994). The questions raised by these observations could only be resolved by serious archaeological study that would extend the ground-breaking, but superficial, research of the pioneer of archaeology in the *Oriente,* the late father Pedro Porras. It was in these circumstances that Stéphen Rostain, whose amassed his vast archaeological experience during his study of raised fields in the Guianas, was recruited by the IFEA to conduct a program of excavations in the Ecuadorian Amazonia.

As circumstances had it, in extending an operation already begun by the IFEA with the Ecuadorian archaeologist Ernesto Salazar, Stéphen Rostain studied mound sites in the Upano valley, rather than the archaeology of the lowlands further east. This orientation was entirely consistent with the specific skills Rostain had gained in analogous sites in the Guianas. Despite my regret in seeing research on the Pastaza basin deferred, it was clear that the excavation of Huapula, which I had the opportunity to observe in situ, was of remarkable interest for gaining a better knowledge of the pre-Columbian past of the foothills of Ecuadorian Amazonia. The Huapula excavation site offered Rostain the opportunity to deepen the treatment of questions surrounding raised-field agriculture and place them in a broader context: nearly 50 years after the pioneering work of William Denevan on the human-modified landscapes of the Llanos de Mojos, he could pursue the study of Amerindian earthworks, providing a comparative scope that the issues regarding this theme had never previously received.

Two traits attract attention in Rostain's work, which are particularly embodied by this book. On the one hand, one notes the extraordinary intellectual curiosity he has demonstrated for everything concerning the cultural history of the lowlands of South America. He is a voracious reader who absorbs everything that concerns this region, first of course archaeology, from the most recent studies to the oldest, but also ethnology, ethnohistory, colonial history and geography. This insatiable erudition has made him a well-rounded scholar, joining those who, like the pioneer tropical Americanists Métraux, Nordenskiöld, or Koch-Grünberg, have exploited all registers of knowledge to render to this ignored continent a little of the historical and cultural depth that it has long lacked. The other aspect of his work that inspires admiration is his willingness to address comprehensively the issue of raised fields not only in the lowlands of South America, but also in other regions of the world where raised-field agriculture was or still is practiced, by exploring all facets of this agricultural technique, not only archaeological but also ethnological, and its consequences for soils, plants, and environments. In short, Stéphen Rostain is not only a remarkable field archaeologist, he is also a comparativist with broad perspectives and a solid erudition which have made him an authority in the still-embryonic field of Amazonian archaeology, and more widely on the theme of sustainable development in tropical forest environments.

This book also has the merit of asking questions that will require long-term collaborative efforts between archaeologists and ethnologists specializing in the Amazon Basin. I will mention only two questions because they are absolutely essential. The first is that of social hierarchy.

The extensive and durable earthworks that Rostain and archaeologists of his generation have highlighted in many regions of Amazonia did not necessarily require a centralized authority capable of mobilizing the workforce in a more or less coercive manner, as was thought by archaeologists influenced by Steward's neoevolutionism, a school that tended to qualify with the term "chiefdom" everything that appeared to manifest a certain degree of coordination. However, this somewhat old-fashioned term does not match much of anything recognizable in what is known about Native American political organization, either contemporary or during the colonial era. Even if everyone now accepts that current societies—most often fluid, headless, and rather scattered in space—probably are not faithful representations of the panorama of pre-colonial Amazonia, and that larger and more densely populated villages were probably common in many parts, this still does not mean that these villages were organized around a centralized authority. In cases where the first colonial chronicles mention chiefs with extensive power, as in the Llanos de Mojos or on the banks of the Amazon, it would be absurd to ignore this aspect; elsewhere, where only material remains allow inferences about social hierarchy, we should proceed with caution.

Thus, the description that Rostain gives of the very structured organization of farmlands in Arauquinoid sites of the Guianas coast is strongly evocative of tropical horticultural societies in the highlands of Papua New Guinea. In the Mount Hagen area, for example, rich volcanic soils allow permanent settlements and human densities of the order of 120 inhabitants per square kilometer, but without centralized chieftaincies (Strathern 1971:231).[1] Nested descent groups and segmental solidarities, strengthened regularly by ceremonial exchanges, are sufficient to ensure the coordination necessary for the very structured organization of village farmlands. It must also be recognized that complex tasks do not necessarily require centralized coordination: the example of the Achuar shows that spacious houses, or large dams for fishing, can be built collectively without the necessity of an overseer directing the work of others. This form of collective intelligence exists when each person is versatile, adapting to each task by observing those around him. It is likely that this kind of spontaneous coordination could often have structured the collective work required for construction of earthworks.

The second issue is that of trade specialization of Amerindian groups, insofar as it can be reconstructed from archaeological remains. Contemporary ethnology highlights that trade exchanges do not follow economic logic in which each group would trade its specialized productions for those of others, but instead are the expression of arrangements for regional integrations that organize the division of labor between groups in such a way that they are forced to exchange their production to be dependent upon each other and thus constitute an area of peace.

The case of the Vaupés, or that of the Xingu, shows that each group has the technical means and resources to produce all objects—hammocks, canoes, pottery, manioc graters, etc.—that it acquires from the other groups in return for its specialized production, but that this permanent movement of exchange, based on a concerted and artificial scarcity, is a powerful factor of integration between neighboring societies whose relations were initially dominated by mutual distrust, or even undisguised hostility.[2] In other words, the diffusion of objects documented by archaeologists is perhaps less a centrifugal movement from a native home, than a centripetal integration at the regional level that circulates, within ethnic confederations, the specialized productions that define for each group a local identity.

On all these issues, the ethnoarchaeological approach is very promising. It is this approach that is implemented by a new generation of archaeologists of the Amazon—Clark Erickson, Michael Heckenberger, Eduardo Neves, Stéphen Rostain, or Denise Schaan come particularly to mind—who know how to combine classical techniques of their discipline with inputs of ethnology, ethnobiology, and ethnohistory to shed new light on the historical trajectory of Amerindian civilization. This book is a glowing testimony to the promise of this approach.

Philippe Descola (translated by Doyle McKey)

Introduction

"So Much Water!"[1]

After five centuries of exploration and colonization, the Amazon maintains its own legends and its particular aura of mystery. When one says "Amazon," wild and fantastic visions come to mind, inspired by literature and other media. The dominant view of the general public regarding ancient inhabitants of the Amazon Basin differs little from that of the European conquerors in the sixteenth century, who reduced the indigenous cultures they encountered to the rank of animal. The result is a deeply flawed image of life among Amazon's first inhabitants. One of the more extreme examples is presented in the *Smithsonian Atlas of the Amazon*, written by ecologists at the Smithsonian Institution (Goulding et al. 2003). Although this work purports to present a panorama of the region, one detects no mention of indigenous peoples, past or present, in the tropical forest. In fact, virtually no works to date present the pre-Columbian past of the Amazon. The most recent works are a short Brazilian publication, written for a general audience, which summarizes the latest findings (Neves 2006), and a book on ecological history (Schaan 2012). Prior to that, the catalog of the "Unknown Amazon" exhibition in London provided a detailed panorama of the issue (McEwan et al. 2001). The well-illustrated "*L'Archéologie de l'Amazonie,*" though published more than 80 years ago, remains a classic, unsubstitutable source (Nordenskiöld 1930).

The ancient populations of indigenous Amazonians are often thought of as small, nomadic groups, surviving as best as possible in a hostile world. This reductive vision of the Amazon and its inhabitants contrasts starkly with the reality of the region's remarkable cultural and biological diversity.

Nomadic, starving, and lost: does this image of the indigenous groups correspond to the reality?

Nomadic? Nomadism was much less frequent during the pre-Columbian period than it is today. After the demographic collapse caused by the disease epidemics that followed the European conquest, many village communities dispersed (although European priests often tried to reassemble these scattered groups in their missions). Many indigenous people today live in semisedentary settlements, and large villages have existed since the pre-Columbian period, such as in the Upper Xingu region (Heckenberger 2005). Futhermore, the first European explorers noted large villages that extended up to 7 kilometers along the Amazon River. Recently, archaeologists have discovered similar pre-Columbian settlements.

Starving? Modern indigenous peoples are often starving, but this is generally due to the impact of colonists on their resources. Indigenous groups prefer to live and eat well, and a slight belly is considered a sign of good health. Unconcerned with food storage, they collect what they need from their fields, the forest, and the river. In contrast, during the pre-Columbian period, there was occasional surplus production and the storing of food reserves.

Lost? Until the 1980s, it was believed that no large society could develop in the "humid desert." Theories of ecological limitations (Meggers 1971), which were based on studies from the natural sciences and applied Western reasoning to the management of the rainforest, presented an Amazonia populated by tribal societies. However, the world's largest rainforest experienced more complex pre-Columbian development than is apparent at first glance. By occupying the fertile alluvial plains, indigenous groups developed a combination of extensive and intensive agricultural systems that facilitated sedentism and increased populations. This expansion resulted in the specialization of tasks and hierarchization, as well as the beginnings of urbanism.

Amazonian archaeology has experienced a tremendous resurgence in the last 30 years. Its current debates address agricultural capacity, chiefdoms, warfare, ecology, demography, and cultural identity. Type-chronologies remain a priority because the stages of settlement are still poorly understood. Nevertheless, ceramics are no longer the sole subject of study; materials such as stone and bone are now analyzed more frequently, as are soils, pollen, and phytoliths. Study of social organization is growing, and Amazonian researchers are increasingly attempting to detect the existence of chiefdoms. As archaeologists have come to believe that humans must be studied in the context of their environment and their impact on the landscape, mostly have integrated geoarchaeology into their research. Historical ecology (Balée and Erickson 2006) is a promising approach that unites multiple disciplines around a shared theme of research. Finally, cultural resources management has been well developed in the last 20 years because of major construction projects, such as the hydroelectric dam at Petit-Saut in French Guiana in the 1990s (Vacher et al. 1998) and the Coari-Manaus gas pipeline project in Brazil (Eduardo Neves, personal communication, 2008).

When I began to work in the region 25 years ago, I could count the number of other Amazon archaeologists on one hand. But today they have become more numerous, particularly in Brazil, where thanks to the dynamic initiated by the late Jim Petersen and by Eduardo Góes Neves, the students and the theses have multiplied. In French Guiana, the second doctoral dissertation in archaeology was just recently defended (Coutet 2010), and three others are in progress. The future of archaeology of the South American lowlands appears bright indeed.

Beyond Nature and Culture

Water is the central element of agriculture. It is essential to the growth of plants, and makes up approximately 80 percent of their fresh weight. Water transports nutrients, promotes the chemical reactions of different life cycles, and maintains cells. A plant will die if it loses 15% to 20% of its water content. Throughout time, farmers have developed systems of water control that allow them to cope with shortfall or excess. Rich in a variety of resources, perpetually or seasonally flooded areas have always attracted humans. However, to cultivate these spaces, they must be managed, adapted, and transformed. The most common method is to create elevated fields and drain the excess water (Figure 1).

Yet, although the basic principle of producing an alternating series of mounds and canals seems simple, the shapes and sizes of these

Figure 1. Raised fields for rice cultivation in Senegal (Pélissier 1966:plate 48).

modifications can be quite varied. The depths, widths, lengths, and course of the ditches can show very notable differences. Furthermore, the farming plots themselves present extremely diverse shapes and sizes. Multiple secondary structures may accompany these sytems (causeways, dikes, reservoirs, etc.). These technological responses reveal a wide range of local adaptations. Raised fields are well adapted to the growth of root crops (manioc, sweet potatoes, and yams), but other plants may be cultivated on them as well. Although raised field agriculture is no longer widely practiced today, it was quite widespread in the past. These structures are frequently found on the coast of Guiana, the remnants of a technique widely used during the pre-Columbian period. They represent archaeological artifacts in their own right (or more precisely, ecofacts).

I have been interested in the pre-Columbian earthworks of the Amazon for more than 20 years, and particularly in the raised fields of Guiana. I use the term "Guiana" rather than the plural "Guianas" because the latter generally refers more strictly to the three coastal Guianas (Guyana, Suriname, and French Guiana), and excludes the adjoining regions in Brazil and Venezuela. Instead, I wish to designate the totality of the area between the Amazon and Orinoco. This demarcation also suggests an island in the true sense, as the region is surrounded by water on all sides (the Atlantic Ocean, the Orinoco River, the Amazon River, the Rio Negro, and the Casiquiare Canal).

My participation in an ongoing interdisciplinary research project focused on the raised fields of Guiana has brought my research back full circle to the initial discovery of these agricultural structures in 1989 (Rostain 1991). This project has combined ecological and anthropological approaches to investigate these landscapes produced by human actions and the forces of nature, yielding striking results (Iriarte et al. 2012; McKey et al. 2010). The seasonally flooded coastal savannas, originally though to be quite homogeneous, are now understood to be heterogeneous products of the combined action of numerous human, animal, and vegetal "ecosystem engineers." The vast pre-Columbian cultural heritage of Guiana is now better understood, and the natural dynamics and processes have been more precisely identified. Pre-Columbian raised field agriculture of Guiana represents a widespread trend throughout Latin America that began with the era of European contact. Humans favored zones that bordered sources of water or were periodically flooded, and they invented many intensive agricultural techniques to expand these areas. It is crucial to study these agricultural methods and the pre-Columbian earthworks, as well as compare them with those created after the European conquest, to better understand the Amazonian landscape.

Three terms are used here to designate the artificial agricultural structures of Guiana:

- **Raised fields** are mounds of diverse shapes and sizes, located in seasonally flooded zones. They have many names in English, Spanish, and French related to local idiosyncracies and subtly different techniques: raised field, ridged field, *campo elevado, camellón, waru-waru, pijal, chinampa, hortillonnage, billon, butte, ado, cavaillon,* etc. I have chosen to use the generic term *raised field* rather than the other names, except for when discussing specific examples.
- **Drained fields** differ most notably by the absence of a raised planting surface. Shallow drainage canals are excavated and their fill is placed on the planting field without creating notable mounding. This technique is also used by the Karinya of the Llanos in the Orinoco (Denevan and Schwerin 1978) and reported at pre-Columbian sites in the Llanos of the Mojos in Bolivia (Denevan 2001).
- Finally, **agricultural beds** are large surfaces, often rectangular, slightly elevated (from 10 to 20 centimeters on average), not necessarily located in zones susceptible to flooding. They can be considered a mixed system, somewhere between raised and drained fields, and the channels that separate the beds generally serve to both drain *and* irrigate. If this technique was used in the pre-Columbian period, no trace of it has yet been discovered. However, various populations have commonly employed this technique in Guiana since the colonial period.

Wetland Agriculture

Guiana, like the whole of the Amazon, is a mosaic of varied landscapes: dense tropical rainforest, *tepuies*, inselbergs, dry savannas, and flooded savannas. These latter habitats, which dominate the low Quaternary plain of the coast, have long been considered unsuitable for agriculture by Europeans.

Representing 6 percent of the global landmass, wetland environments have often been ignored by large agricultural movements. However, populations around the world have invented many ingenious techniques to cultivate the "drowned earth" (Mollard and Walter 2008). Many scholars believe that the inhabitants of these zones were driven from their original territory by more powerful groups and have taken refuge in these zones, which are a priori unfavorable. However, Amazonian ethnohistory and archaeology suggest that this was often not the case in the past. To the contrary, it appears that some communities deliberately chose to settle these flooded environments because they were rich in sources of protein. While these wetland regions impose certain constraints on their inhabitants, they also offer an extremely diverse range of resources. Thus, adaptation to this type of landscape is often easier than it initially appears. In other cases, humans invest a great deal of labor to modify the landscape to serve their needs.

A study incorporating aerial imagery from Google Earth highlights the importance of agricultural modifications to wetland landscapes across the globe (Figure 2).

It is outside the scope of this work to describe all such sites throughout the world. However, it is useful to discuss the world's oldest known raised fields. In western Papua New Guinea, archaeological excavations have revealed an independent agricultural development in the swamps of the Kuk region (1,560 meters above sea level [asl]) that extended over 116 hectares and that were cultivated for close to 7,000 years (Denham 2005; Denham et al. 2003, 2004; Neumann 2003). After progressively cutting back the forest over several hundred years, humans constructed circular raised fields in the flooded areas, where they cultivated taro (*Colocasia esculenta*) and bananas (*Musa* sp.), and, more recently, yams (*Dioscorea* sp.) and probably sugar cane (*Saccharum*), almost continually for close to 7,000 years (6,950–6,440 years calibrated before present [cal BP]).

In New Caledonia, the Kanak society of the pre-Contact era (AD 1000–1774) profoundly transformed their landscape by contructing habitation mounds, causeways, canals, and elongated raised fields (Figure 3). These features strongly structured the landscape and indicate social hierarchization. Kanak territorial space is fundamentally inseparable from the community roots that forged the unity and identity of the nation (Dotte et al. 2005).

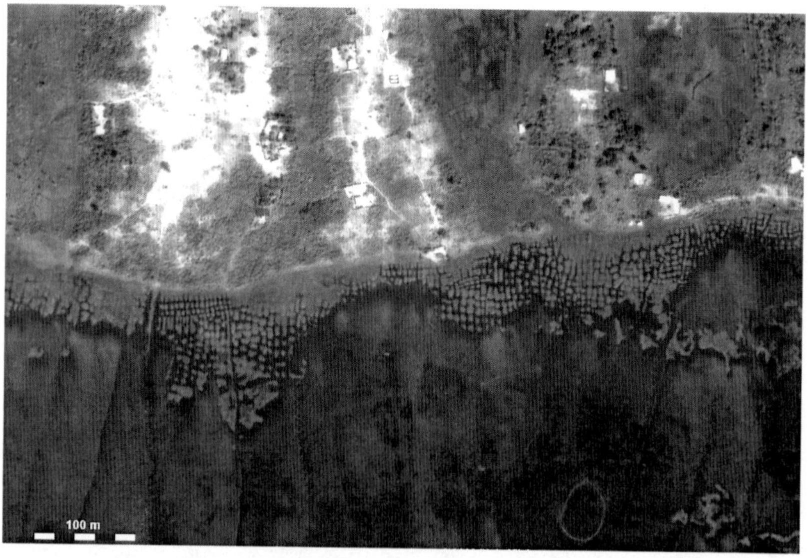

Figure 2. Relic raised fields near fishponds (*acadja*) on Nokoué Lake, near Cotonou, Benin (adapted from Google Earth; altitude 827 meters). © 2012 Google. Image © 2012 DigitalGlobe. Map Data © 2012 AND.

Raised field techniques exist in many different variations throughout the world, and are practiced in both temperate and tropical climates. While their shape and dimensions may vary significantly, the basic principle is universal: cultivate above the water in a zone that is flooded or prone to flooding.

The most elaborate technique of raised fields is found in the Mexican *chinampas*[2] (Armillas 1971). *Chinampas* can be considered a *hortillonnage*, comparable to that practiced in Amiens, Picardie, for close to 2,000 years.[3] *Chinampas* were constructed during the pre-Columbian period in the Xochimilco, Chalco, Xaltocán, Texcoco, and Zumpango lakes of the Basin of Mexico (Figure 4). The basic principle is to create arable land within a lake. As there is no information available regarding how they were built, it was long assumed that they began as rafts made from tree branches and earth that was progressively heaped with successive layers of sediment.[4] However, as *chinampas* continue to be used and maintained today, we can observe their construction. The floating aquatic vegetation present on the lake is cut into strips, and this resulting vegetal material is then piled on the shallow lake bottom until it rises just above the water's surface.

Figure 3. Raised fields and residential mound in the lowlands of Tiwaka, New Caledonia (photo IGN, 1991).

This layer is then covered with mud, and canes are planted around the plot and allowed to root into the lake bottom. A stabilizing "fence" is created by weaving branches between the canes. In this way, the agricultural beds are irrigated by the capture of water through the roots of the plants. The soil is renewed regularly by adding new layers of mud to increase its fertility. Today, *chinampas* are used to cultivate maize and a variety of other species for subsistence and medicinal uses, as well as flowers (Mollard and Walter 2008; Whitmore and Turner II 2001).

The origins of the *chinampas* remain uncertain and are probably quite ancient, but by AD 650, the populations displaced by the collapse of Teotihuacan began to modify the edges of Lake Chalco, then at a very high level, with the use of elongated raised fields and *chinampas* (Niederberger Betton 1987). Several centuries later, the Aztecs developed this technique to a remarkable level beginning at AD 1400, responding to the enormous needs of a growing population in and around Tenochtitlán. It is estimated that shortly before the arrival of the Spanish, 100,000 people were fed by the production of close to 9,000 hectares of *chinampas* (Niederberger Betton 1987).

Other wetland agricultural systems similar to *chinampas* are used around the world (Mollard and Walter 2008). The *hortillons* of the

Figure 4. Xochimilco *chinampas*, Mexico (photo by Stéphen Rostain).

Amiens region are plots 50 to 100 meters long by 80 to 150 centimeters wide, separated by canals that are permanently flooded. The *maa* of Micronesia are small mounds built in the marshes to cultivate taro, a plant that requires very humid conditions. Similarly, the floating garden is a raft built from a natural tangle of roots and aquatic plants or a trellis of branches, upon which a layer of earth is spread.[5] They are cultivated in lakes in Asia, Africa, and Latin America.

Agricultural responses in flooded or seasonally flooded environments vary greatly across the world and demonstrate the virtuosity and the ingenuity of their creators. When one does not cultivate wetland plants, many different kinds of artificial soils are created above the flood level. Humans understand how to establish a subtle relationship between earth and water for farming. Unfortunately, we tend to forget these agrarian techniques in our modern era, and they are often replaced by mechanized agriculture.

In contrast to the Andean highlands, irrigation is not usually necessary in the Amazon; the principal problem is drainage. Nevertheless, it is sometimes useful to channel water, particularly in the dry season. Furthermore, several periods of drought have affected the largest tropical rainforest for several thousands of years (Meggers 1994, 2007), which may have required irrigated farming techniques. The drastic drop in the level of the Amazon in October 2005 caused by exceptional drought (a direct consequence of El Niño), reminds us that this rainforest is fragile and sensitive to variations in precipitation.

Historical Ecology of the Guiana

The coast of Guiana provides an excellent example of the pre-Columbian preference for wetland environments, which goes back thousands of years. In the 7th millennium BC, the Amazon's first sedentary fisher-gatherers, the Alaka culture, colonized the western coast of Guiana along the left bank of the Orinoco Delta, constructing enormous shell middens called *sambaquís*, upon which their villages were founded. These middens are evidence of the importance of marine shellfish for these populations, who appear to have been attracted by richness of the fauna and flora. Much later, other communities chose to inhabit this type of landscape, which, although seemingly forbidding, is in reality a source of abundant and diverse resources. Pre-Columbian settlers also carefully transformed this landscape through time to increase its resource potential. These profound modifications have left indelible marks on the ecosystem that are nevertheless difficult to detect using classic approaches because they are often subtle and sometimes hidden by more recent changes. Innovative research methods, such as aerial archaeology or interdisciplinary research designs, are essential for understanding the past transformations of this region.

Since the discovery of raised fields in French Guiana in the late 1980s, science has made tremendous progress in many domains, such as the study of sediments and paleoethnobotanical remains, and the development of new dating techniques, allowing a more nuanced understanding of the archaeological remains and the environment. It was difficult to imagine 20 years ago that we could identify the plants that were cultivated on raised fields and those consumed in the settlements. Yet, the study of phytoliths, pollen, and starch grains has made these interpretations possible. Similarly, increased research on raised fields throughout Latin America offers a much wider corpus of comparative data, particularly in Bolivia and Colombia. Growing emphasis on interdisciplinary studies has also played an important role. In the 1980s, convincing a group of researchers to share data across their different demains posed an incredible challenge. Fortunately, collaborative projects that involve a diverse number of specialists organized around a common research problem are widely accepted today.

This book represents a synthesis of 25 years of research in the archaeology of Guiana. The data were collected from fieldwork, museums, archaeological storehouses, and libraries from different countries. The book is organized in five chapters:

- Chapter 1 describes other areas of raised fields in South America, principally those located in Amazonia, on the Caribbean coast, on the Pacific coast, and the Andean highlands. Other major agricultural techniques in the Amazon are also discussed. Finally, different types of pre-Columbian Amazonian landscape modifications are presented. These are principally artificial earthen mounds with domestic, funerary, and/or ceremonial functions, often associated with causeways, raised fields, canals and reservoirs, or fortified ditched sites.
- Chapter 2 presents the environmental context from the perspective of human occupation and land use. Guiana developed culturally in concentric zones, beginning at its center. Initially, hunter-gatherers spread across the interior savannas. Later, farming groups began practicing agriculture in the surrounding forests. Finally, stratified societies colonized the margins bordering the ocean and rivers. What advantages did these landscapes and biotopes offer humans? What challenges did they pose?
- Chapter 3 presents the history of human occupation in Guiana from its origins to the arrival of Europeans. Hunter-gatherers, fisher-gatherers, cultivators, and farmers occupied the region at various moments over several millennia, leaving indelible marks on the

landscape. Recent archaeological discoveries have improved our understanding of the pre-Columbian chronocultural panorama.

- Chapter 4 discusses different aspects of pre-Columbian agricultural modifications in Guiana. More than 10 years ago, William Denevan insisted on the importance of these agricultural features, which "*need more study*" (2001:227). Since that publication, two major research projects have improved our understanding of these agricultural constructions considerably (Figure 5).
- Chapter 5 describes the use and history of Guiana's coastal savannas since the conquest: their abandonment by native peoples, their adoption by colonists and later by the Creoles, and finally, their rediscovery by modern farmers. Modifications are varied, and sometimes significant, although they do not equal the initial investment of effort by their pre-Columbian founders. The chapter closes with a discussion of the historical reasons for the destruction or preservation of these indigenous structures today.

Today, we acknowledge that the landscapes upon which we live are as much the result of ancient human actions as they are of natural processes (Frenzel 1979; Leroyer et al. 2008). Only through multidisciplinary collaboration are we able to understand our biotope in all of its dimensions. The archaeologist plays a critical role in interpreting

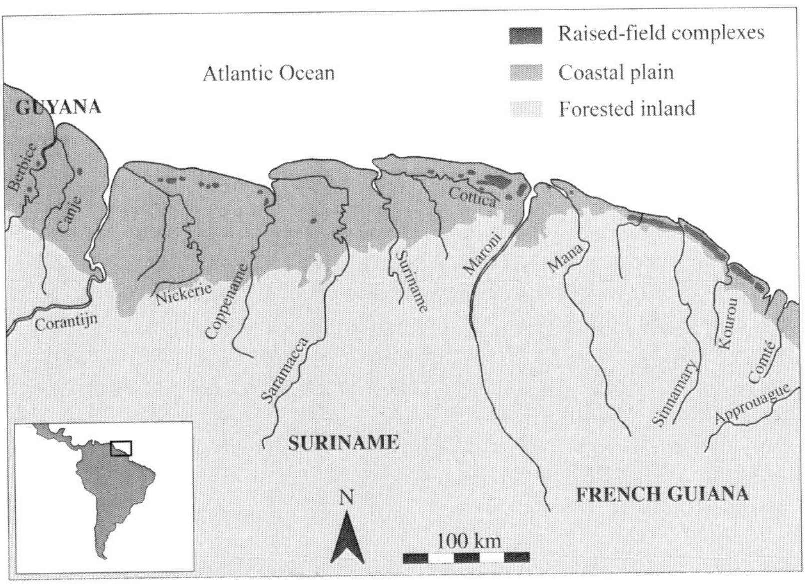

Figure 5. Map of raised-field areas along the Guiana coast (drawing by Stéphen Rostain).

the evidence by bridging studies of past human behavior with studies of the environment. Due to the richness of its physical setting and its pre-Columbian heritage, the Amazon offers a unique field of investigation (which remains in great part still unexplored) for such an endeavor.

Chapter 1

Indigenous Agricultural *Savoir Faire*

South American pre-Columbian populations greatly modified their landscapes, particularly those areas that were otherwise not well adapted to agriculture. In a few cases, irrigation was the primary concern, but more commonly the goal was to control excess water through drainage systems. In both the highlands and the lowlands, the most typical form of raised field is the elongated raised field, but the internal organization of these complexes can vary considerably. The principal difference among these systems is whether they articulate with rivers, standing water, or periodically flooded spaces.

The study of raised fields in Latin America is characterized by 50 years of research and turbulent polemics. Dennis Puleston (1977, 1978) initially believed that the human-constructed raised fields in the Maya region of the Yucatan Peninsula were natural in origin and resulted from long-term processes of alternating swelling and shrinkage cycles of the clay substrate (i.e., gilgai). It was similarly necessary to demonstrate in French Guiana that raised fields were intentional modifications to the landscape rather than gilgai (Rostain 1991). Nevertheless, it is still possible to read that South American raised fields are formed by the actions of rodents or termites, or even processes linked to the growth of shrubs (Meggers 2003). Yet, the distinction between structures of natural and anthropogenic origins is not always clear-cut because some mounds result from a combination of the two factors (Mayle et al. 2007). Whereas some landscapes are clearly abandoned raised fields, other seemingly regular landscapes exist that have formed through natural processes, without human influence (Rietkerk et al. 2004; Rietkerk and van de Koppel 2008).

Indigenous groups also developed agricultural techniques to complement raised field systems, such as agroforestry, gardens, and other more elaborate techniques. Three of the latter are presented here, either due to their widespread usage (e.g., slash-and-burn techniques), or due to the technological expertise they require (e.g., *várzea* and *terra preta* agriculture). Finally, it is important to recall that indigenous peoples were not only skilled farmers, but also experienced engineers of nonagricultural

Figure 6. Pre-Columbian earthworks in South America with the exception of *sambaquís* and artificial mounds south of Capricorn Tropic (drawing by Stéphen Rostain on map by Graphi-Ogre).

constructions, specifically the earthen or clay mounds upon which they constructed their villages, buried their dead, or carried out ceremonies. These are often associated with both raised structures (causeways or dikes), or structures in relief (canals or reservoirs). Other sites are surrounded by impressive ditches. In sum, the earthworks developed by indigenous groups (particularly those of the Amazon) represent a remarkable example of monumental architecture that we are only beginning to understand.

South American Raised Fields

Raised fields are found in a number of Latin American countries, particularly those north of the equator (Figure 6). While most are located in the seasonally flooded lowlands, some are situated in the Andean highlands. These are generally elongated raised fields or, less commonly, rounded or square mounds, such as those that are found in Guiana. Most of these structures were built during the first millennium of the Common Era, but some were constructed as early as 1000 BC.

The geographer William Devenan was the first to study these structures in depth in the 1960s, developing a typology of their shapes and arrangements (Denevan 1966, 1970). Since then, a variety of additional research has been carried out on raised fields, ranging from simple descriptive inventories to detailed analysis of their soils or archaeobotanical remains, or even their construction and experimental cultivation of the fields.

Amazonian Raised Fields

The remains of pre-Columbian raised fields are preserved in two regions of Amazonia, the Llanos of Venezuela and Bolivia.

At the end of the 1960s, raised fields were detected in the Llanos of Apure, Venezuela (Zucchi and Denevan 1979). Study of the Caño Ventosidad complex revealed elongated raised fields arranged in pairs, which are divided by a canal that runs perpendicular to the river, in a zone covering a total of 1,550 hectares. Almost all of the 500 to 525 structures are 1,000 meters long (some are 2,000 meters in length) and average 15.5 meters wide. An intermediate-sized canal is 4.4 meters wide. The average space between paired mounds is 48.8 meters. The height of the mounds is from 25 to 50 centimeters above the level of the canal. However, the canals have been partially filled in with colluvium since their abandonment, and were likely originally at least 1 meter deeper (Zucchi and Denevan 1979). Preliminary surveys have not found habitation sites associated with these raised fields, but pre-Columbian settlements are often buried and not visible from the surface (as has been observed in French Guiana). The closest habitation was localized on a small, forested island a dozen kilometers away. It has been attributed to the

El Choque culture, dated from AD 1200 to 1400 and affiliated with the Arawak tradition.

Other raised-field complexes have been identified in this region, but there are likely many that have been buried by the significant sedimentation on the savannas (Denevan 2001). The Hato La Calzada complex, 100 kilometers to the west of Caño Ventosidad, consists of 10.3 hectares of elongated raised fields. The La Tigra site contains 35 hectares of drained fields and a corresponding network of canals with a combined length of 3,800 meters (Spencer et al. 1994). Although these systems appear to have been produced by a variety of local cultures, ceramic evidence indicates that all of these groups shared Arawak roots.

Since first described by Erland Nordenkiöld ([1916] 2009), the raised fields of the Llanos of Mojos, Bolivia, have drawn the attention of researchers (unfortunately, often at the expense of the pre-Columbian settlements themselves). The structures are located in a wide, marshy savanna that covers close to 90,000 square kilometers (Denevan 1966). Estimates of the extent of the agricultural features themselves vary. Some authors base their calculations only on the distribution of the fields themselves, whereas others consider the size of the complexes as a whole. Finally, some researchers measure the total area within which agricultural features are distributed. After studying aerial photography of close to 35,000 structures and a dozen large complexes of small mounds, William Denevan (2001) calculated that the total coverage was at least 6,000 hectares, of which 2,600 hectares represent actual cultivated surface. Assuming a population density of 22 persons per cultivated hectare (based on experiments carried out in Mexico, Ecuador, and Bolivia), the same author calculates a total of 57,200 inhabitants if all 2,600 hectares were under cultivation at the same time (which is unlikely). A different estimate was made based on the manpower necessary to construct and maintain the raised fields compared with their carrying capacity. The conclusion is that a social group of 20 to 100 individuals could have constructed a raised-field complex in a short period of time to meet their subsistence needs (Walker 2004).

Four principal types of pre-Columbian structures are distinguished: low rectangular platforms, narrow elongated raised fields, small mounds, and drained fields (Denevan 1966, 2001). Platforms measure up to 335 meters long, 5 to 20 meters wide, and 0.15 to 0.6 meters high. They can be either isolated or found in groups of several hundred (Figure 7). Elongated raised fields are 6 to 300 meters long, 1.5 to 6 meters wide, and 0.3 to 0.9 meters high. They are generally associated with other pre-Columbian constructions, such as artificial settlement mounds, causeways, and the latter two types of fields. The small mounds are 1.2 to 1.5 meters long, 0.6 to 0.9 meters wide, and 0.3 to 1.0 meters high. Finally, drained fields are rectangular parcels surrounded on three to four sides by ditches that range from 60 to 120 meters long.

From AD 500 to 1400, local stratified societies carried out significant construction projects: residential mounds, raised fields, causeways,

Figure 7. Raised fields of the Llanos of Mojos, Bolivia (photo by John Walker).

canals, and artificial reservoirs. Dating of the raised fields yields a range that covers the first 1,200 years of the Common Era (Erickson 1995). More recent dating indicates their use was concentrated in two major periods: the large San Juan structures are dated between AD 410 and 665, and the smaller El Cerro structures are from AD 1281 to 1644 (Walker 2004).

Experimental fields were constructed and cultivated in the region, producing 25 tons of manioc and 2 tons of maize per hectare (Erickson 1995, 2006a). However, such high rates have been tempered by the hypothesis that raised fields were not an alternative technique but an adaptation to specific environmental pressure (Lombardo et al. 2011). Among the experimental attempts, one of the more unusual was that of incorporating water hyacinth (*Eichornia crassipes*, Pontederiaceae) as a source of mulch in a raised-field complex (Barba et al. 2003; Saavedra Arteaga 2006).

Coastal Raised Fields

Numerous raised-field complexes are found along the Pacific coast and the northern Columbian coast. Those of the Sinú Basin and San Jorge are the best studied from the Caribbean coast of Colombia (Figure 8). In 1965, thousands of elongated raised fields were discovered along the

Figure 8. Raised fields of the San Jorge Basin, Colombia (Colección Museo del Oro, Banco de la República, Bogotá, Colombia; photo by Sebastián Shrimpf).

Cauca River and the San Jorge River (a tributary of the Magdalena), as well as along the Sinú River (Parsons and Bowen 1966). However, study of these features did not begin until 15 years later (Plazas and Falchetti de Saenz 1981). The hydraulic systems of the San Jorge cover approximately 90,000 hectares (Plazas and Falchetti de Saenz 1987), and consist of a series of parallel canals that run perpendicular to natural streams, forming a checkerboard pattern. The peculiarity of this raised-field complex is that it appears to be associated with systems of not only drainage, but also irrigation. The fields are raised to be above the flood level, but are also oriented perpendicular to the flow of water to allow irrigation. However, experiments conducted in Bolivia show that local hydrological and pedological conditions had an important impact on raised field technique, and that the reasons for building raised fields can be complex (Lombardo et al. 2011).

Radiocarbon dating of samples from the canals indicates a range between 810 and 330 BC. However, this type of agriculture was intensified more recently, and then abandoned sometime around 1000 AD (Plazas et al. 1993). Even if these mounds were built and used over several centuries, they would have provided for the needs of a substantial population.

Several artificial settlement or funerary platforms were built close to the raised fields. Their occupation begins around AD 300 and is followed by a decline around AD 900 (Plazas et al. 1993). The territory of these communities was organized along waterways, with the residential and funerary mounds intercalated between the elongated raised fields.

In the Calima Basin in Colombia, raised fields are dated between AD 100 and 1200 (Bray et al. 1987). To the south, along the border between Colombia and Ecuador, La Tolita populations constructed multiple raised-field complexes around 1000 BC, each of which covers more than 8 to 10 hectares (Valdez 2008a). The elongated raised fields, rectangular or L- or Z-shaped, measure 5 to 25 meters long and 6 meters wide. Their average height is 0.4 to 0.5 meters above the water (Figure 9). Streams are also modified by dikes and deviations, and the largest canals empty into small artificial reservoirs. Causeways structure the landscape. They measure 3 meters wide on average, up to 1 meter high, and can extend for more than 2 kilometers (Valdez 2008a).

Most of the cultural remains excavated in the marshes correspond to the post–La Tolita occupations, but the construction of raised fields goes back at least to AD 800. A layer of volcanic ash covers the last archaeological level throughout the region, up to 3 to 5 centimeters thick in the flooded zones, where it is best preserved. It probably corresponds to the Quilotoa eruption dated to 810 years BP, thus placing the abandonment of these sites in the thirteenth century AD. No subsequent pre-Columbian occupation of the area has been detected (Valdez 2008a).

Figure 9. Raised fields of the El Indio complex, La Tola, Ecuador (photo by Francisco Valdez).

Some experiments with maize cultivation (and with some secondary plants) have been carried out on the L-shaped, pre-Columbian, elongated raised fields at the mouth of the Santiago-Cayapas River (Cadudal 2007). These fields measure 60 centimeters high, and cover a parcel of 0.15 hectares. The maize planted in the canals grew more quickly than that planted on the raised fields. The harvest, which was carried out 72 days after planting, yielded approximately 20 kilograms per hectare, which is two times *less* than the minimum obtained on modern fields using chemical fertilizer. However, it is important to emphasize that this technique permits six years of continuous cultivation through a system of crop rotation and does not require fertilizers. In contrast, fields on sand spits must be fallowed after only two to three years of cultivation (Cadudal 2007). Although these experiments were only carried out over a short period of time (3 months) and should be complemented by additional longer-term studies, they nevertheless provide preliminary data regarding these agricultural techniques.

One of the largest complexes of raised fields (covering more than 50,000 hectares) is found in the Guayas Basin, east of Guayaquil, in coastal Ecuador (Denevan and Mathewson 1983; Parsons 1969; Parsons and Denevan 1967). Unfortunately, they have been almost completely destroyed by modern rice farming, but several research projects have worked to document the elongated raised fields before their complete disappearance (Denevan 2001). One of these projects involved building and cultivating experimental raised fields in 1983, and the high production figures presented in Chapter 4 are based upon these results (Muse and Quintero 1987).

The oldest raised fields at Guayas date back between 400 and 200 BC. Their period of greatest expansion occurred between AD 400 and 600, but they continued to be used until AD 1650 (Stemper 1993). These structures are associated with other earthworks such as large, artificial mounds and long, linear platforms that likely served residential, ceremonial, or funerary functions (Meggers 1966). Maize phytoliths were recovered during excavations of the raised fields (Stemper 1993).

Raised fields have been detected on the desert coast of central Peru, at the mouth of the Casma River, which is periodically flooded and marshy. There are only 240 hectares still visible. The elongated raised fields measure 80 meters long, 2 to 3 meters wide, and 0.4 to 1.3 meters high (Moore 1988). They have been attributed to the Chimú culture, which dates from AD 1300 to 1470 (Pozorski and Pozorski 1982).

Raised-field complexes have been described close to Lake Budi, on the central coast of Chile, and also to the north, in the Purén and Lumaco Valleys (Dillehay et al. 2007). The elongated raised fields of Lake Budi, arranged parallel to one another and perpendicular to the shore, measure from 18 to 20 meters wide, 1.2 to 1.7 meters high, and up to 1,400 meters long. They make up 80 percent of the surface, with the intermediary canals occupying the remainder. The mounds of the Purén Valley, distributed in 2- to 5-hectare complexes, measure 8 to 15 meters long, 3 meters wide, and 0.5 to 1.5 meters high. Macrorests of maize (*Zea mays*), madia (*Madia sativa* [Asteraceae], probably for medicinal use), quinoa (*Chenopodium quinua*, Chenopodiaceae), and an unidentified fruit (Poaceae) have been identified. The paleosols of the raised fields of Lake Budi have been dated to AD 1145 ± 110 and AD 1210 to 1320, while those at Purén produced dates of AD 1120 to 1420 (Dillehay et al. 2007). These raised fields are assumed to be associated with the arrival of the complex societies that occupied the region between AD 1000 and 1500.

Andean Raised Fields

Raised fields are found in the highland valleys of the Andes, where their principal function is drainage. The presence of water around the fields also serves to dampen the variability between daytime and nighttime temperatures, a function that differentiates these structures from those of the lowlands. In the Cayambe region of Ecuador, during nighttime frosts, the temperature of the raised field surfaces is up to 1.3 degrees Celsius higher than that of adjacent nonmodified ground surfaces (Knapp 1991). Whereas some raised fields are constructed in flood-prone zones, others are located on the edges of lakes rather than rivers, where there is a need for both irrigation and drainage. Thus, although the form of highland raised fields may resemble those of the lowlands, they clearly serve a distinct agricultural purpose.

Numerous elongated raised fields were noted in the savanna of Bogotá (2,600 meters asl) before they were partially destroyed by urbanization.

They were initially attributed to the hierarchical Chibcha society and dated to AD 1100 (Broadbent 1964, 1987). Continued research has expanded the chronological range from 1100 BC to AD 1800, with a period of intensification between AD 400 and 900. Maize and bean pollen have been detected in these raised fields (Boada Rivas 2006).

Many raised-field complexes have been detected in the highland valleys of northern Ecuador: San Pablo, Cayambe, Paquiestancia, San José de Minas, San Pedro, Peguche, Angochagua, San Leonardo, and possibly at La Libertad and San Isidro. Initial estimates suggested that they covered 2,278 hectares (Gondard and López 1983), but more recent work indicates their coverage is even more extensive. Raised-field complexes were detected south of Quito, but were subsequently destroyed (Knapp and Ryder 1983). In Quito itself, small structures uncovered during construction projects may date as far back as 300 BC (Villalba and Alvarado 1998).

The single raised-field site discovered at Cayambe (Figure 10) covers 1,237 hectares (Gondard and López 1983). Phytolith analysis revealed the presence of maize, but no detailed data are presented in the study (Villalba Sevilla 2007). The raised fields are associated with Andean chiefdoms that also constructed high earthen ceremonial platforms with ramps (Caillavet 1983, 2006). There is evidence of the use of manure as fertilizer on the raised fields of Zuleta (Wilson et al. 2002).

Figure 10. Raised fields of Cayambe, Andes, Ecuador (photo by Manuel Sigüenza).

The first raised fields would have been constructed around AD 600. An eruption of Quilatoa subsequently buried them under a layer of ash, giving the terminus post quem date for their use as AD 1280 (Mothes 1998). Colonial archives from the sixteenth and seventeenth centuries repeatedly mention the use of *camellones* (elongated raised fields) in this region (Caillavet 2006).

There are around 122,000 hectares of elongated raised fields around Lake Titicaca, located at 3,800 meters asl, in the northwest of Puno, Peru, and the border with Bolivia (Denevan 2001; Erickson 2000b). With a height of 15 to 125 centimeters, they are organized into several patterns (Figure 11) (Denevan 2001). Those of the "checkerboard" variety measure 2 to 40 meters long and 5 to 20 meters wide. Others are ringed by an irregular embankment that protects them from flooding. On slopes, very elongated raised fields measure on average 8.7 meters wide and are up to 500 meters long. Arranged in rows, they are sometimes almost square with an average length of 40 to 70 meters and a width of 15 to 35 meters. They occasionally are oriented perpendicular to a stream or other waterway. The most ancient raised fields, located north of Lake Titicaca, date to 1000 BC. After an interruption of construction between AD 330 and 1000, the technique was continued until around AD 1450 (Darch 1983; Denevan et al. 1987; Erickson 1987). The structures located south of the lake were slightly later, with dates

Figure 11. Raised fields of Lake Titicaca, Peru (photo by Stéphen Rostain).

between 800 and 200 BC, and were used until around AD 1000 (Kolata 1991).

Along with the complexes from the Llanos of Mojos, the raised fields at Lake Titicaca are those that have most attracted the attention of researchers, who have constructed experimental raised fields in collaboration with local populations (Erickson 2003a; Kendall and Rodríguez 2009; Morris 2004; Swartley 2002; de la Torre and Burga 1986). These farming experiments have produced varied estimates of annual productivity, ranging from 1,500 to 6,000 kilograms/hectare (Kolata 1991), 8,000 to 16,000 kilograms/hectare (Erickson 1993), or 8,000 to 21,000 kilograms/hectare (Kolata 1996). As a result, estimates of the human population that raised fields supported range widely from 215,000 to 229,3000 individuals (Denevan 2001), and should be considered with caution. The experiments have also inspired a renaissance of this agricultural practice, supported by the Peruvian and Bolivian governments. However, modern raised fields have sometimes been abandoned by local farmers for sociocultural or economic reasons (Bandy 2005; Erickson 2003b; Swartley 2002).

OTHER AGRICULTURAL TECHNIQUES

While shifting cultivation is the dominant practice today in the Amazon, diverse farming techniques were developed during the pre-Columbian era, such as those practiced on *terra preta*, the alluvial plains, or raised fields. Excluding the latter, the remaining principal pre-Columbian agricultural techniques are slash-and-burn, *várzea* farming, and *terra preta* farming.

Slash and Burn

Today, most native groups practice slash-and-burn agriculture, a technique of shifting cultivation that involves opening a patch in the forest (called *abattis* in French Guiana) through tree-cutting and the use of fire. The ash serves as a temporary fertilizer for the poor soil. The field also can only be cultivated for a few years, after which it must be subjected to a long fallow period. The decision to abandon the field often results primarily from the increased growth of weeds and, consequently, the increased time and labor required to remove them (Doyle McKey, personal communication, 2010). The fertility of rainforest soils is based on the local biomass (90 percent) and only slightly on the surface litter. For that reason, it is necessary to temporarily enrich the soil with ash and, at the same time, to break up the soil with fire (Renoux et al. 2003). Fallowing does not necessarily imply the complete abandonment of the parcel, as collection of useful plants and planting of desired species continues even in the fallow period. Indeed, as many modern indigenous groups have (re)adopted sedentary lifeways, it is often necessary to maximize the use of the environment around the settlement, even if it means overexploitation and a reduction in the length of the fallow period (which provokes a decrease in yields) (Renoux et al. 2003). Thus, prolonged sedentism becomes detrimental over the long term and counterproductive to slash-and-burn agriculture.

Although the western Amazon is characterized by a diverse number of cultivated crops, manioc (*Manihot esculenta*) is the principal crop of the Amazonian slash-and-burn field in the present day. This root crop presents a number of advantages, such as reproduction through cuttings, tolerance of poor soils, resistance to predators and parasites, and its ability to survive for more than two years in the ground (Arroyo-Kalin 2008, 2010; McKey and Beckerman 1993). Among indigenous groups of the Amazon, one crop normally dominates cultivation (often bitter manioc, maize, sweet potato, or banana), but a number of secondary groups are cultivated simultaneously. For example, the Wayana of French Guiana manage a territory of approximately 2 hectares per family, in which a parcel of at least 0.5 hectares is opened each year. Approximately 20 species are cultivated there over a two-year cycle (Fleury 2000; Hurault 1965). The Wayampi of the upper Oyapock cultivate 35 species (Grenand 1981). In the equatorial Amazon, the majority of the Achuar fields contain 62 different species (Descola 1986). Fruit trees are planted around the house and the surroundings of the village are cultivated. For example, on one 3-kilometer section of the path that sets out from the large village of Kayapó, 185 trees have been planted (representing 15 different species), as well as close to 1,500 medicinal plants and 5,500 edible plants (Posey 1985).

Shifting cultivation exists in a number of close variations, such as short fallowing, rain feeding, or agroforestry (Denevan 2001). The fundamental link among all of these techniques is the regrowth of forest cover (Grenand 1996). Furthermore, the exploitation of plants can be very simple, such as favoring the growth of certain nondomesticated species. Groups of mobile Paleoindian hunter-gatherers no doubt used similar practices. They could encourage the regrowth of edible plants along the routes they regularly followed, would have been the first to study their local flora, and may even have understood the properties of these plants better than later farming populations.

It is frequently accepted that slash-and-burn agriculture was the earliest and most common agricultural practice used by indigenous people before the European conquest, beginning around 6,000 to 5,000 years BP. Maize recovered from Lake Ayauchi in the Equatorial Amazon, dated to 5,300 years BP, demonstrates the antiquity of the domestication of plants in this region (Bush et al. 1989). Wood charcoal recovered near San Carlos, on the Upper Negro of Venezuela, dates from 4000 BC to AD 1750 and probably represents remains from agricultural burning and wildfires (Clark and Uhl 1987). Nevertheless, assumptions for the primacy of the slash-and-burn technique must be regarded with caution. Slash-and-burn agriculture may actually be relatively recent (Denevan 1992b). Indeed, it is extremely difficult and time consuming to cut trees with a stone axe (Carneiro 1979; Pétrequin and Pétrequin 1993). My own experiences with cutting trees using a stone axe in French Guiana in the 1980s demonstrate the need for great patience. Fire would have been a useful addition to carry out the task.

The blades of stone axes in Guiana vary greatly in their shape and size (Boomert 1979; Rostain and Wack 1987), but two principal methods of hafting are recognized: (1) the blade inserted into the handle, or (2) the blade simply attached to the haft with ligatures and/or an adhesive (Rostain 1990). Studies of hafting and microwear on the blades have not provided evidence of the use of adzes; the axe dominates (Rostain 1994a). Given the large diameter of certain tree species and their high density in the forest, it is astonishing that indigenous groups did not adopt the adze, which is extremely useful in such an environment (see the case of Irian-Jaya [Pétrequin and Pétrequin 1993]). Slash-and-burn agriculture has become associated with European contact because it became the predominant technique among the indigenous groups after their numbers decreased drastically due to demographic collapse and the dissolution of populations provoked by the arrival of the Europeans. The introduction of the metal axe, one of the principal bartered objects since the beginning of the colonial period, was the main factor that allowed increased forest clearance. Furthermore, iron axes played an essential role in the Jesuits' evangelization of indigenous groups (Métraux 1959). Moreover, Father Chantre y Herrera wrote in the seventeenth century "it is rare that it be Divine Reasons—that the Indians little heed—that attract them to our mission. They settle there for very practical reasons. . . . We could not accomplish anything without the axes we distribute" (Chantre y Herrera [1767] 1901:265, 276).

Várzea Agriculture

The second pre-Columbian technique is the floodplain farming practiced along the Amazon and the Orinoco during the period of low water. The alluvial plains, or *várzea*, occupy only a small portion of the Amazon Basin (5 to 10 percent), which is between 154,400 and 300,000 square kilometers along the length of the river and its principal tributaries (Denevan 2001). Scientists have referred to a dichotomy between *várzea* and *terra firme* to distinguish between fertile and poor soils in the Amazon (with the *várzea* favored for farming). However, the indigenous conception of these soils is much more nuanced and involves specific agricultural technological responses developed for each of these landscapes.

The raised fields used to farm the *várzea* represent an ingenious adaptation to a wetland environment. *Várzea* agriculture is carried out on beaches, islands, and riverbanks during low-water periods. Annual flooding deposits fertile sediments that permit semiannual harvests. The low *várzea* is inundated by 3 to 4 meters of water for a large part of the year (Goulding et al. 2003; Wittmann et al. 2004). On the Middle Amazon, the beaches can only be cultivated for a maximum of four months, and the higher portions of the *várzea* for approximately eight months (Padoch et al. 1999). Despite these restrictions, the annually renewed soils of the *várzea* are among the most fertile of the Amazon. It is likely that this agricultural method was crucial for the emergence of the numerous stratified societies that still populated the *várzea* of the Amazon and the Orinoco at the time of contact. *Várzea* agriculture permitted the production of surplus that would have been exchanged or stored in pits for consumption during the season when the floodplain could not be farmed.

In the eighteenth century, the Otomac, inhabitants of the savannas between the Orinoco and the Apure of Venezuela, cultivated along the edges of lagunas when these areas were not flooded (Gumilla [1745] 1963). However, cultivating the alluvial plain is not risk-free. It was probably necessary to complement the cultivation of very fertile, seasonally flooded areas by farming the less-productive nonflooded banks of the *terra firme* (Denevan 1996). Although the period of high water on the more fertile soil is largely predictable, sudden and unpredictable rises in the river (where the water may rise several meters in a single night) can destroy the harvest. Supplementary planting on the less productive, but non–seasonally flooded, *terra firme* would have helped to mitigate this risk.

Várzea agriculture is still used today in the Amazon, notably by the Achuar of Pastaza, Ecuador (Descola 1986), and the Shipibo of Peru (Hiraoka 1986). It was adopted by colonists who used it in Colombia as a complement to slash-and-burn farming (Augusto Oyuela-Caycedo, personal communication, 2007). It is also the technique used in the Palcazu Valley of Peru by colonists (McClain and Cossío 2003). In Brazil, a large part of the Amazonian alluvial plain is cultivated during the low-water season (Grenand and Grenand 1993), particularly with the construction of *lejas*, planting beds used for drainage (Eduardo Neves, personal communication, 2010). In Ecuador, the shoals are farmed at the mouth of the Pastaza, which descends from the Andes towards the Amazon. Many more examples can be cited for most of the Amazonian floodplains.

Terra Preta Agriculture

Another type of farming was practiced on the fertile black anthropogenic soils called *terra preta* (Balée 2010b; Glaser 2007). Tremendous interest in this technique has emerged in the last 20 years (Arroyo-Kalin 2008; Eden et al. 1984; Glaser and Woods 2004; Kern 1988; Lehmann et al. 2003; Petersen et al. 2001; Woods 1995; and Woods et al. 2009, to name a few). The excellent fertility of the *terra preta* soils is due to the presence of wood charcoal, organic material, and anthropogenic micro-remains (Arroyo-Kalin 2008). Slow processes of microbial degradation of the charcoal produces carboxyl groups with a weak negative charge. The tiny fragments of charcoal are thus chemical analogs to clay particles of the types typically found in other types of fertile soils, with a strong capacity to retain cations (because of the weak negative charge of the oxygen atoms). This capacity is stronger than that of clay minerals typically found in leached tropical soils, such as kaolinite (Glaser et al. 2001). Experimentation has shown that charcoal can increase the fertility up to 880 percent (Glaser et al. 2001).

Terra preta expanses measure between 0.1 and 5 hectares, although some exceed 300 hectares (Petersen et al. 2001). Their average thickness is from 40 to 60 centimeters, but can measure up to more than 2 meters. *Terra preta* expanses may cover up to 10 percent of the Amazon (Glaser and Woods 2004). These expanses have been mapped along the length of the Amazon River, but it is likely that they are much more widely distributed. For example, they have been detected in Colombia, along the Middle Caquetá (Eden et al. 1984; Hiraoka and Mora 2001; Mora et al. 1991). Samples of black soil collected from anthropogenic levels of excavations at Huapula, Ecuador, have revealed phosphate levels comparable with those of *terra preta* around modern indigenous villages (Petersen et al. 2001). The *terra preta* of the Hatahara site on the Middle Amazon has been dated to 360 BC (Figure 12), and it may be even more ancient at other sites, up to 4,000 years BP (Neves 2006). They are most common throughout the Amazon between AD 450 and 950 (Petersen et al. 2001).

After years of polemics, detailed soil analysis has allowed scientists to reach a consensus on the dark soil's origins. *Terra preta* results from intensive and long human occupation of a single location, while *terra mulata*, a lighter colored soil, develops in areas where repetitive fire-intensive cultivation of slash-and-burn fields took place (Arroyo-Kalin 2012; Denevan 2004). In any case, these dark soils reflect the intensity of human occupation during the pre-Columbian period in the Amazon Valley. Analyses have also demonstrated that maize and manioc were their principal crops, but at least 30 other economic species have also been identified (Arroyo-Kalin 2008; Fraser 2010; Fraser and Clément 2008). *Terra firme*, representing close to 90 percent of the Amazon, has long been considered a limiting factor on the development and sedentism of human societies, as well as on their demographic growth (Meggers 1971, 1996; Steward 1948). However, this view does not consider the

Figure 12. A thick level of *terra preta* in the Hatahara site, Middle Amazon, with the archaeologist Eduardo Neves near funerary urns (photo by Manuel Arroyo-Kalin).

remarkable agricultural strategies developed by indigenous farmers, such as *terra preta* and *terra mulata* (Lehmann et al. 2006).

Pre-Columbian Earthworks

Fifteen years ago, Claude Lévi-Strauss, the father of Structuralism, argued that the pre-Columbian Amazon was significantly more culturally diverse and complex than previously assumed, an assertion that has since been supported by archaeological research: ". . . the illusion was firmly established that the present conditions of the Indian communities prolonged those which had existed at the moment of their discovery. . . . [T]he Amazonian rainforest is not uniformly as 'primary' as we like to say. . . . The Amazon could be the cradle from which sprang the Andean civilizations" (Lévi-Strauss 1994:12–13). Indeed, the discovery of immense settlements and the existence of complex societies by the new generation of archaeologist working in the Amazon since the 1980s has provoked radical revision of our assumptions about pre-Columbian peoples. We became aware of the size and density of the earthworks from the Llanos of Mojos, Bolivia (Erickson 1993). Similarly, work at the Marajoara mounds of Marajó Island revealed that the existence of complex societies in the region was not the result of an ancient migration from the Andes (Meggers and Evans 1957), but rather the product of a long-term, local development (Roosevelt 1991). However, if recent investigations revealed the importance of landscape management by pre-Columbian societies, it must be stressed that such human impact has been heterogeneous and did not affect the entire Amazonia, but only specific areas (Barlow et al. 2011; Bush and Silman 2007).

A few years later, Eduardo Neves and James Petersen discovered and studied several enormous sites along the Middle Amazon, which often contained settlement mounds (Lima et al. 2006; Neves and Petersen 2006).

At the same time, in the Upper Xingu of Brazil, Michael Heckenberger (2005; Heckenberger et al. 2008) excavated the site of Kuhikugu, whose major development occurred around AD 1510. With a surface area of 50 hectares, it is the largest residential site of the region. Its organization is similar to that of the nearby modern Kayapó villages, consisting of a ring-shaped settlement oriented around an immense central plaza. The large communal houses are located on the periphery and are protected by two rows of ditches. Very wide roads originate from the village in a radial pattern to connect it with secondary sites. A variety of fairly large earthworks mark the landscape: docks, bridge foundations, etc. The territory as a whole is organized in a system of principal centers and secondary settlements, indicating the existence of a stratified society. Today, the indigenous populations of the Upper Xingu still live in this type of large, ring-shaped village, connected to one another by wide, straight roads.

Pre-Columbian earthworks are varied, but share three principal functions. The first is for settlement on artificial mounds, which includes funerary functions (burials have frequently been discovered in these residential mounds) as well as ceremonial uses. Certain mounds were dedicated solely to funerary functions, such as the San Jorge Basin of Colombia (Plazas and Falchetti de Saenz 1981). The second function is for defense, through ditches excavated on the periphery of settlements. The third function, discussed in detail previously, is agriculture on raised fields. Such earthworks have been recorded in different locations throughout the Amazon.

Settlement Mounds

Many complexes of artificial settlement mounds are found throughout South America, the most famous of which are the *sambaquí* shell middens of the Atlantic coast. Earthen mounds are relatively common on the Pacific coastal plains, from Colombia to Peru, and on the Brazilian and Uruguyan coasts. In the Amazon, it is very probable that there are complexes that remain undiscovered. Others of modest dimensions are still being studied, such as the low elevations that appear in some sites of the Middle Amazon (Eduardo Neves, personal communication, 2010).

The mounds of Marajó Island have been known since the end of the nineteenth century (Derby 1879, 1898; Hartt 1871b) and those of the Llanos of Mojos since the early twentieth century (Nordenskiöld 1930). Several investigations of these sites have been carried out. Settlement mounds on the Andean piedmont and the Llanos of Venezuela have been recorded and studied more recently (Porras 1987; Zucchi and Denevan 1979).

The most ancient complexes are those of the Upano Valley in Ecuador, whereas the others have been dated to after the beginning of the Common Era. These societies of earth builders, perhaps organized hierarchically, reached their apogee around AD 1000, and some persisted until European contact.

The Upano Valley of Ecuador

Pre-Columbian populations began very early on to construct earthen mounds for habitation in the Amazon. The most ancient earthworks that have been identified are located in the upper equatorial Amazon, on the Andean piedmont. Originating at the base of the Sangay Volcano, the Upano River flows parallel to the Andes and is bordered by terraces. More than 2,000 mounds are arranged in groups along the summit of these terraces. Very rapidly, between 500 and 200 BC, communities of the Upano culture built settlement platforms from clay sediment, known locally as *tolas* (Rostain 1999a). Dozens of mound complexes were discovered in this valley at the end of the 1970s, and the largest site was the subject of archaeological excavation (Porras 1987). New research was carried out on different complexes as a part of the Sangay-Upano Project (Rostain 1999b; Salazar 2008) and Río Blanco Project (Rostain 2006a) between 1996 and 2001.

The inhabitants of the Upano Valley constructed platforms and level and low plazas, and dug paths (Figure 13). In several cases, they integrated the slope or other natural relief into the mounds' construction to minimize the transport of fill. Because the sediment becomes muddy and slippery in the rainy season, they burned the tops of the mounds to create a compact and hardened floor. The mounds are oval or rectangular, and in a few cases, L- or T-shaped, or very rarely circular. The upper surface is leveled, or slightly convex. At the top, they measure between 10 and 50 meters long, 3 to 10 meters wide, with an average height of 2 to 10 meters.

These sites are organized according to a repeating spatial model. The basic pattern consists of a central plaza (generally square) that is enclosed by platforms. This model exists in several variations, such as

Figure 13. Artificial earth mound complex at the Domono site, Upano Valley, Ecuador (photo by Stéphen Rostain).

the presence of a central mound or six peripheral mounds. Dug paths, which serve to reduce the natural slope to the neighboring stream, are often associated with these complexes. Most of these mound complexes were residential. Other complexes, such as the notably straight peripheral mounds, must have served an additional function that is not yet understood (possibly a secondary ceremonial use).

Huapula (previously called "Sangay" by Pedro Porras in 1987) is the largest site of the Middle Upano, extending over 70 hectares. It is located on a terrace that overlooks the Huapula River to the east, and the Junguna to the west (which flows into the Upano farther downstream). Hualapa is made up of an enormous group of complexes and smaller entities built on relatively flat areas and separated by deep ravines. The principal site is composed of approximately 30 of these mound complexes, with a central group of particularly large and high mounds, which cover close to 50,000 square meters. The lateral groups, to the northwest and the southeast, are smaller and lower. The site became famous 20 years ago when Pedro Porras interpreted the arrangement of these mounds as the stylized representation, viewed from above, of a man copulating with a jaguar (Porras 1987). New mapping of the site has more accurately recorded the locations of the mounds, demonstrating the absence of such imagery, but nevertheless presenting evidence of a predefined spatial model (Rostain 1997).

Three straight dug paths run parallel to one another across this space, the largest of which is 1,200 meters long, and whose widths measure up to 13 meters, with a current depth of 3 meters (they have been partially filled in by erosion) (Salazar 2008). They are connected perpendicularly by secondary paths. Bordered by a rim on one or two sides, they extend for dozens or hundreds of meters, with a width of 2.5 to 10 meters and a depth greater than 3 meters (Rostain 1999a).

During the Sangay-Upano Project, three different areas of the Sangay site were excavated: a midden without a mound close to the Junguna Ravine, a large two-level mound at the center of the site, and the group from Complex XI that is the furthest from the center. This last group covers approximately 3,500 square meters on the banks of the Huapula River. It is organized according to the typical spatial model of the Upano sites, with a central plaza divided in two by a platform and surrounded by six mounds. Several areas on top of the mounds and in the plaza were excavated level by level (Rostain 1999b). The stratigraphy of these mounds shows that an initial occupation, attributed to the Sangay culture (700–500 BC), occurred directly on the sterile level. This occupation was subsequently covered by construction fill, and the top of the mound was burned to harden the surface and facilitate settlement construction by the Upano (500 BC–AD 400/600) until the massive eruption of Sangay, which deposited a tremendous amount of ash in the valley and destroyed the settlement. The volcano is located 35 kilometers northwest of the site. At Huapula, the layer of ash is more than 30 centimeters thick and decreases progressively to the south before finally disappearing at the village of Sucúa 60 kilometers south of the volcano. After this catastrophe, the Upano communities permanently abandoned the valley and migrated south to the Ucayali River in Peru. A few centuries later, between AD 800 and 1200, the Upano Basin was settled by Huapula populations, who took advantage of the existing mounds by building their houses upon them. The recent excavations in the Huapula site have revealed that they appear to be the ancestors of the Jivaro populations that currently occupy the region (Rostain 2006a, 2012).

The Marajoara Territory of Brazil

Marajó is an enormous island of 49,000 square kilometers at the mouth of the Amazon in Brazil. The eastern half of the island is flooded in the rainy season, making settlement difficult. As a result, pre-Columbian populations constructed elongated mounds, known locally as *tesos*, which were most often grouped parallel to a stream or scattered around Lake Ararí (Meggers and Evans 1957). It seems that the builders took advantage of Late Pleistocene and Holocene raised sandy formations to construct the mounds, reducing the amount of labor and materials necessary (Rossetti et al. 2009). Indeed, some of the mounds contain up to 630,000 cubic meters of earth, which would have represented a considerable amount of work.

The mounds have an average height of 3 to 7 meters, but can be as high as 20 meters (Figure 14). They generally measure from 600 to 2,500 square meters, with some measuring close to 900,000 square meters. They were constructed by Marajoara communities between AD 450 and 1350 (Meggers and Evans 1957; Roosevelt 1991; Schaan 1997, 2000). Residential mounds are differentiated from ceremonial ones, which were probably used by the elite (Schaan 2004, 2012). They are arranged in large groups. If all of the mounds at Camutins (the largest site) were occupied at the same time, this would have represented between 28 and 43 large communal houses, or close to 2,400 inhabitants (Roosevelt 1991).

Fishing reservoirs appear to have been constructed close to the mounds as reserves for the low-water season (Schaan 2004). The construction of pools and dams in the river would have allowed for plentiful fishing, providing sufficient protein for demographic growth. It has even been suggested that the Marajoara societies depended on the intensive exploitation of aquatic resources, freeing them from the need to intensify their agricultural production to serve the needs of a large population (Schaan 2008).

Figure 14. Artificial earth mound on Marajó Island, Brazil (photo by Serge Guiraud).

The abundance of remains from ceremonial and funerary activities in the largest mounds suggests they served as political and ceremonial centers that dominated the smaller sites. Like other contemporary Amazonian cultures that had adopted Polychrome and Arauquinoid traditions, the Marajoaran societies are considered to represent the emergence of social complexity in the Amazonian lowlands.

The Llanos of Barinas and Apure, Venezuela

Different types of pre-Columbian earthworks, often associated with raised fields, have been recorded over the last 30 years in the Llanos of Barinas and Apure, Venezuela, but they have rarely been the objects of archaeological research.

Nonetheless, three projects were carried out in the 1980s and 1990s on the sites of the state of Barinas, with two doctoral theses completed (Garson 1980; Gassón 1998). In addition, Elsa Redmond and Charles Spencer organized surveys of 103 sites and excavations from 1983 to 1988, covering both the Andean piedmont and the *llanos* (Redmond and Spencer 1998, 2007; Spencer et al. 1994). One of the most important achievements of this research is a regional comparative study of two different environments (seasonally flooded savannas and the foothills of the Andes) that analyzes the network of interactions between several groups. The *llanos* sites are made up of artificial earthen mounds. The piedmont sites are recognized by the presence of petroglyphs on outcroppings. They do not contain mounded architecture. Elsa Redmond and Charles Spencer classify sites of the upper *llanos* into four categories: regional centers that cover more than 10 hectares with mounds, plazas, and causeways that connect two other types of constructions; secondary centers, from 5 to 10 hectares, with artificial mounds; tertiary centers, from 1 to 5 hectares, without architecture; and raised-field complexes.

A chronological sequence spanning close to 1,500 years has been defined for the region (Redmond and Spencer 2007). It is separated into two major phases: Curbatí-Gaván (AD 300–1000) and Caño Seco-Chuponal (AD 1000–1850).

The six Curbatí sites, principally located on the piedmont, vary from 2 to 8 hectares and do not contain earthen architecture, but petroglyphs have been observed. It appears that a regional centralized power was established during this period.

The 38 Gaván residential sites and the two raised-field complexes are located in the *llanos*, the majority in the non–seasonally flooded zones. Chiefdom level organization is not evident in the Early Gaván phase (AD 300–550), but earthen architecture of the Late Gaván phase (AD 550–1000) shows three levels of settlement hierarchy, suggesting the appearance of a chiefdom and a strong increase in population. The 38 sites vary from 1 to 33 hectares, 28 of which measure less than 5 hectares. The largest, El Gaván, is composed of impressive earthworks with numerous mounds of different dimensions, including one with an access ramp, a surrounding embankment, and causeways that lead to the secondary sites. The regional communication network permitted exchange and the control of the territory by the chiefs. These roads would have simplified the distribution of agricultural surplus and facilitated military movement. Evidence of warfare has been recovered from excavations in the center of El Gaván. Alignments of carbonized postholes are apparently the remains of a defensive palisade that was destroyed at the time of the site's abandonment between AD 760 and 900.

Most of the 27 Caño Seco sites are located in the piedmont. Although their surface area ranges from 1 to 25 hectares, most are less than 4 hectares. These villages were obviously constructed on the best agricultural land. Again, the increase in the number and size of these sites appears to indicate population growth. Two levels of settlement hierarchy are distinguished and centralized power is assumed, but it is difficult to verify the existence of one chiefdom. As in the preceding period, the cultivation of maize and manioc is confirmed.

The 11 Chupoal colonial sites occupy riverbanks on the upland *llanos*. They generally do not exceed 7 hectares, but one site covers 26 hectares and another 80 hectares. European artifacts are sometimes associated with indigenous artifacts.

The Llanos of Mojos, Bolivia

Complex societies occupied the Llanos of Mojos, Bolivia, constructing settlement mounds, raised fields, causeways, canals, and fishing reservoirs (Erickson 2008). Clark Erickson created a detailed description of the mounds (known locally as *lomas*) (2000c). They are almost always arranged in groups and distributed along a stream, around a flat space, or around a large marsh. They are oval or rounded, sometimes stretching in an arc around the edges of a flooded area, or in even more complex arrangements (doubled or multiplied). Erickson distinguishes four types of mounds according to their size and location. There are 200 to 300 large mounds that cover several hectares each and measure between 3 and 8 meters high. Several thousand medium-sized mounds cover 1 to 2 hectares each and have a height of 1 to 3 meters. There are even larger numbers of small mounds of less than 1 hectare and 1 meter high that served as house foundations. Finally, there are long, rectangular mounds that are organized parallel to one another around a plaza, distant from residential settlement, suggesting possible ceremonial functions.

Although the remains found on most of these mounds attest to a domestic function, urns containing human remains have been frequently recovered as well, indicating they also served a funerary function. A small mound may be constructed close to the center, on top of larger mounds. These mound complexes often figure into a network of causeways that cross marshy areas. Although raised fields are generally associated with mounds, they are never associated with the largest mounds (Erickson 2000c).

The Mojos *llanos* have has been profoundly transformed by human action, of which the mounds represent only one of the numerous types of constructions. For example, the Loma Salvatierra Mound, measuring 8 meters high and occupied from AD 500 to 1400, is surrounded by an embankment that circumscribes a zone of 20 hectares around the mound (Lombardo and Prümers 2010; Prümers 2004). The settlement is separated from the cemetery, and its surroundings are organized with a network of canals, ditches, circular reservoirs, small mounds, and causeways.

Ditched Sites

Besides the artificial mounds, often associated with paths and raised fields, as well as reservoirs and fishponds, the other notable earthworks found in the Amazon are the peripheral ditches. To date, they have been found in the east of Guiana and the Upper Amazon.

In French Guiana, east of Suriname and the eastern edges of Amapá, more than 40 sites with peripheral ditches have been reported. The Saramaka call them "crowned mountains" (Rostain 1994a). These are generally small hills, close to a river, ringed at the summit by a ditch 3 to 15 meters wide and approximately 1 meter deep, generally filled in one, two, or three places to allow crossing. They measure on average 100 meters in diameter. Some reach up to 800 meters long, 8 meters wide, and 5 meters deep. In one case, the ditch was carved directly into the rock (Éric Gassies, personal communication, 2009). Sometimes, particularly on the high banks of the river, the ditch only closes a hillfort. Little archaeological work has been carried out on these structures. These sites are particularly numerous in the interior of French Guiana and are rare in Amapá and Suriname. They are dated from the beginning of the Common Era to the early colonial period (sixteenth century).

In Brazil, on the right bank of the Middle Oyapock and at the height of the village of Camopi, sherds tempered with coarse-grained quartz (some of which are painted) have been collected at Grand Carapana, a hill site surrounded by an oval ditch (Petitjean Roget 1991).

To the northeast of Suriname, the peripheral ditch of Pondo Kreek-2 was excavated in 1980 (Versteeg 1981). The oval ditch has a diameter of 120 meters by 95 meters, and measures on average 4.2 meters wide, with an original depth of 3 meters. Surface collections yielded 264 quartz-tempered sherds (four of which were painted, or 1.5 percent of the sample), and a quartz polishing stone for ceramics. Three test trenches cut across the ditch produced very little archaeological material

(Aad Versteeg, personal communication, 1994). A piece of wood charcoal, collected at 175 to 180 centimeters below the base of the ditch, was dated to 1180 ± 70 years BP (Versteeg 1981). The exact function of these sites is unknown, but it seems likely that they were fortified villages or ceremonial sites. The thinness of the anthropogenic level and the low percentage of phosphates recorded at Pondo Kreek-2 (Versteeg 1981), as well as the rarity of the archaeological material discovered, do not suggest a long occupation or a permanent settlement.

The ditched hill of Crique Yaou on the Upper Maroni of French Guiana, excavated in 1992, revealed two archaeological levels and furnished a number of ceramic sherds. The two carbon 14 (^{14}C) dates obtained on wood charcoal range between 200 BC and AD 130 (Mazière and Mazière 1993). Dating carried out at other ditched hill sites throughout French Guiana have returned ages between the beginning of the Common Era and European contact (Éric Gassies, personal communication, 2009; Aujoulat et al. 1997).

The oral tradition of the Wayampi of Oyapock attributes these sites to the ancient Karane, who inhabited villages with defensive structures: "throughout the region, there were many Kalana. They constructed villages that they protected with ditches two meters wide and one meter deep. Stakes were planted at the bottom. Those who did not know fell in and died. There are still remains of these ditches around the mountains close to Camopi and elsewhere" (Captain Norbert cited by P. Grenand 1982:270). Moreover, at the end of the eighteenth century, on the Inirinda tributary of the Upper Orinoco, the Puinave (warriors who practiced cannibalism) lived in fortified villages surrounded by ditches filled with poisoned stakes (Dreyfus 1992). At this time, the Wayana villages of French Guiana also presented elaborate defensive systems (Leblond 1813). However, these data are insufficient to prove the military use of pre-Columbian ditched sites. Thus, if the defensive function cannot be discarded, it is more probable that they had a ceremonial use.

In the savannas of the state of Baures, Bolivia, and Acre, Brazil, more than 360 sites with peripheral ditches have been noted (Erickson et al. 2008; Pärssinen and Korpisaari 2003; Schaan et al. 2008). These are large, enclosed fortifications that are either round, oval, quadrangular, or complex, measuring 90 to 320 meters in diameter (Figure 15). Circular and square moats can be associated a short distance from the complex, or even form an elaborate system. This type of site sometimes has been improperly identified as a geoglyph; however, they do not present any clear or recognizable motifs. Although the first ditch was discovered in 1977, the recording and study of these structures only truly began in 2000 with the work of Martti Pärssinen and colleagues (2009) on the Brazilian side and those of Clark Erickson on the Bolivian side.

Figure 15. Ditches at the Fazenda Colorada site, Acre State, Brazil (photo by Martti Pärssinen).

The function of these sites has been originally considered as defensive. The ditch, protected by a slope constructed from its own extracted fill, would have served as a water reservoir. In the thirteenth and fourteenth centuries, a drought in the Andes provoked climatic changes in the Upper Amazon that contributed to migrations, demographic pressure, and warfare (Abbott et al. 1997; Binford et al. 1997; Bray 1995;

DeMenocal 2001; Nunn 2000; Pärssinen and Korpisaari 2003). Ditched sites could be evidence of conflict on the piedmont of the Andes before contact. Archives indicate the existence of villages with palisades and surrounded by ditches during the colonial period (Pärssinen and Korpisaari 2003). However, ^{14}C dates indicate that most of the geometric ditches were dug between 200 cal BC and cal AD 900 (Saunaluoma and Schaan 2012; Schaan 2012). Moreover, the arrangement of some complexes suggests they served a ritual function; recent investigations tend to show that these sites were monumental public spaces reserved for ceremonial purposes (Saunaluoma and Schaan 2012; Schaan 2012).

When considering this overview of Amazonian earthworks, an astonishing variety of structures and landscape modifications come to light. Today, archaeologists and earth science specialists accept that these are pre-Columbian accomplishments and not natural formations (although these do exist in some cases). These constructions suggest the organization of communal labor, cooperative or under the authority of a specialist, and a centralized power. Archaeology has revealed indications of stratified pre-Columbian societies at many points across the South American lowlands. Their appearance is generally dated to around AD 500 to 600, peaking around AD 1000. They begin to decline by AD 1200 and disappear around the time of European contact.

The analysis of pre-Columbian earthen structures indicates a reality that differs considerably from the traditional image of an Amazon inhabited by small, semisedentary tribes, the legacy of an entrenched Eurocentric perspective. Although the pre-Columbian history of the Amazon is not written in books and there are no stone temples, the indigenous peoples of the tropical forest nevertheless inscribed their annals in the earth.

Chapter 2

Humans and Environment: A Happy Marriage

In the Introduction, Guiana has been presented as an island. Why speak of Guiana Island? Early on, chroniclers distinguished Guiana from the rest of the Amazon (Figure 16). In the middle of the seventeenth century, François de Pagan (1655) recognized a coherent ensemble between the Orinoco and the Amazon. At the end of the eighteenth century, Miguel Marmión wrote:

> The Province of Guiana in all its extent, made up of the possessions of the Portuguese, the French, and the Dutch, is more or less one thousand leagues in circumference (5500 km), from the mouth of the Orinoco up to San Carlos in the junction of the Casiquiare with the Rio Negro, following its waters downstream to those of the Amazon, which flows into the Atlantic sea; and turning towards the West, following the coast to the mouth of the Orinoco, of which the last of Sotavento flows into the Triste Gulf across from Port-of-Spain of Trinidad Island (Marmión in Arellano Moreno 1964:438).

Lope de Aguirre was the first European to explore the Casiquiare Canal in 1561. Nevertheless, westerners did not travel regularly through this geographic anomaly until much later, and its existence was previously more understood through myth than direct observation (Alès and Pouyllau 1992). Bordered on all sides by the Atlantic Ocean, the Amazon, the Orinoco, and the Rio Negro, it was not difficult to envisage the egg of Guiana, located between the latter two basins, separated from its Amazonian womb, creating an island of this region that spans 1.8 million square kilometers. What phenomenon or magic allowed a canal, unique in the world, to link two watersheds? This distributary channel of the Upper Orinoco flows over 355 kilometers before it empties into the Rio Negro (Georgescu et al. 2008), called Rio Guainía in Venezuela, to the level of the village of San Carlos. There, the Orinoco loses 12 percent of its waters during low-water periods and 20 percent during floods. Charles de La Condamine was the first scholar to travel along the Casiquiare Canal in 1800 and confirm its existence. A few years later, Alexander von Humboldt descended the canal and created a map.

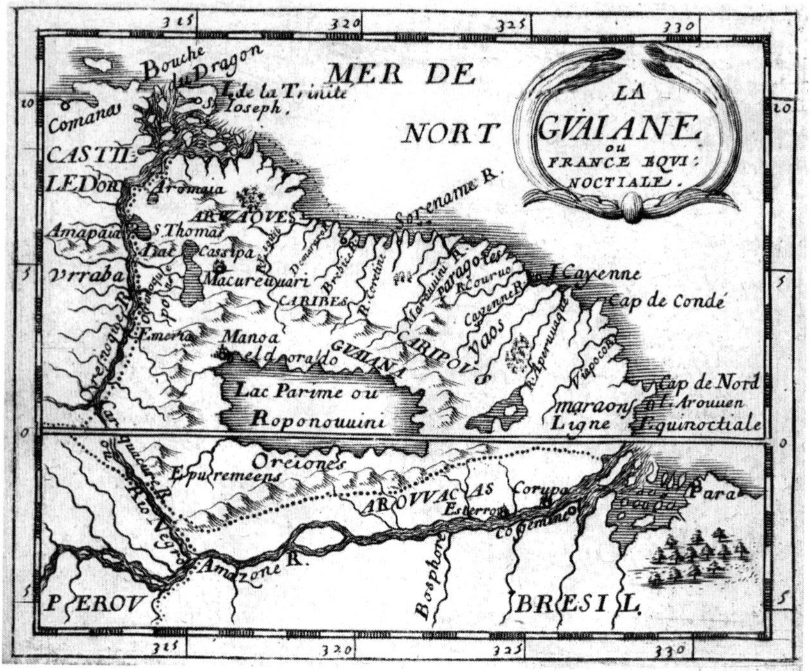

Figure 16. "*La Guaiane.*" Guiana Island closed by the Casiquiare Canal on eighteenth-century map (drawing by Pierre Duval, 1640).

Nonetheless, this breach in the Orinoco continued to attract interest for its seemingly artificial aspect on maps. Having surveyed the region, particularly down the Casiquiare, in 1913, Theodore Koch-Grünberg provided some explanations. In a letter to Erland Norkenskiöld, he responded to the hypothesis that indigenous peoples had modified the first few northern kilometers of the canal (as well as a few small, seasonal streams) (Nordenskiöld [1916] 2009). According to Koch-Grünberg, the upper reaches of the canal were previously a much straighter, steep-sided channel. Flooding had expanded the channel and lowered the angle of its banks. At the time of his visit, it was still difficult to see the entrance to the canal. It was only revealed when his boat was suddenly drawn into the narrow gulley by the current. Only a few kilometers downstream, the stream widened and began to form meanders, receiving small tributaries (one of which may have been built by indigenous peoples).

The German explorer finished the letter to his Swedish colleague with great enthusiasm for this line of research: "Your idea that these canals are perhaps artificial opens an entirely new historic question that would be very interesting to pursue further. It remains remarkable that these canals occur in an Arawak cultural hearth" (Nordenskiöld [1916] 2009:223).

Nevertheless, the original inhabitants of the region were very conscious of the particularity of this waterway that isolated Guiana when they described it to the German geographer Alexander von Humboldt as running in two directions. The sacred aspect of the place was also strongly rooted among the indigenous populations because

> The country is desert, a mysterious terror reigns over this region, and the Indians never venture beyond Mavaca. They tell that, farther along, the inhabitants are cannibals, that they watch for travelers from high above in the trees to shower down poisoned arrows, that invulnerable in their fall, more apes than men, they fall on their prey to devour it. Flames escape from the earth and this land is seeded with horrors (Huber 1887:416).

Since its discovery, Guiana has been both a surprising and misunderstood reality and an overflowing utopia. It is thus "doubly separated from the continent by the Casiquiare Canal (which made it an island) and by the axe that cut Raleigh head (who made it a dream)" (Lézy 2003:14). In 1596, Walter Raleigh, his eyes shining with the gold dust of El Dorado, invented the "Rich kingdom of Manoa" at the center of Guiana, "virgin land that has never been pillaged, cultivated, nor exploited, the face of its land has not been degraded, nor has its virtue or its salt been spoiled by smoke" (Raleigh [1596] 1993).

Another famous myth was that of Parimé Lake, bordered by the extraordinary city of Manoa and represented on maps until the eighteenth century, when Charles de La Condamine ([1745] 1981) destroyed this ghost. There, the formidable Amazons supposedly extracted clay from the lake to fashion their green, frog-shaped pendants (the *muiraquitãs*), which became hard as stone once taken from the water. Chimerical lake, city of gold, cannibals, and fierce female warriors make up a story more fantastic than any Hollywood screenwriter could dream up. Many have fed and fueled the myth. The historian and English scribe of the Amazon, John Hemming (1978, 1987, 1990, 2003, 2008, etc.), created his own legend of the wealthy gentleman, scattering allusions throughout his works that sustain the myth. The gold prospectors who today lay French Guiana bare of its earth and subsoil are only the latest participants in the frantic hunt for golden riches.

This Guiana, sometimes called "the nameless land" or, paradoxically, the land of a thousand names—Guiana, Caribane, The Wild Coast, Eldorado, New Andalousia, or Equinoctial Land—presents many truths. The most amusing is surely that of an island that sprang forth from the Amazon only after the Casiquiare Canal freed it and gave it life. Indeed, the Guiana shield resulted from an independent evolution over 120 million years, which produced a geographic region whose physical, as well as cultural, characteristics distinguish it from the more recent western Amazon (Hammond 2005).

What does this emancipated child look like today (Figure 17)?

GENERAL CHARACTERISTICS

The equatorial climate of the region presents a dry and a wet season, as well as a fairly constant temperature, with an annual average of 26 degrees C. The climate of Guiana is characterized by two wind regimes:

- The Southwest trade winds, originating from the Saint Helen anticyclone, produce a relatively dry wind due to loss of part of their humidity over northern Brazil. The winds result in a dry season from August to October, during which the average rainfall is less than 2 centimeters per month (Cayenne meteorological station).
- The Northeast trade winds, originating from the Azores anticyclone, bring the humid flux directly from the ocean. They result in a rainy season from November to July that interrupts the "Little March Summer," particularly for the western portion of the coast.

The Northern Brazilian Current, and the Guiana Current, which are produced by the trade winds, direct the sediments of the Amazon northwest along the length of its shores. The coastal route by boat from Guiana to the mouth of the Amazon is therefore much slower compared with travel in the other direction, as demonstrated by Antoine Biet's (1664) account of a Kali'na maritime expedition in 1652. The Kali'na left from

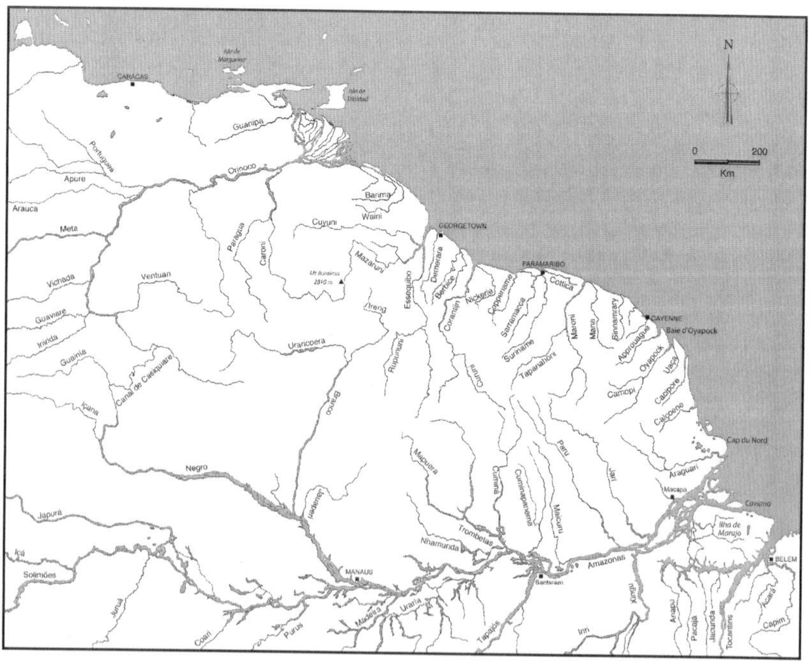

Figure 17. Modern map of the Guiana Island (drawing by Stéphen Rostain).

Île de Cayenne heading towards the Amapá Grande River (*Maricary*) in Amapá, where they were to meet up with the Arikare (*Racalets*). The voyage required two weeks to arrive but only five days for the return trip, where the winds helped, making it possible to use sails.

Despite abundant rainfall (36.74 centimeters per year between 1961 and 1990 at the meteorological station, and 29.74 centimeters at the Cayenne station, for the same period), the hierarchy of the hydrological network is fairly weak, except for the large waterways such as the Araguari, the Oyapock, the Maroni, the Corentijn, or the Essequibo, among others. In contrast, the density of small valleys and minor streams (called "creeks") is very high. These waterways either empty into major rivers or into the ocean (functioning as tidal channels), or they drain into marshes, where combined with rainfall, they increase the depth of the permanent standing water on the low plain.

The velocity of current of the rivers decreases significantly as they approach the coastal band, where numerous meanders exist. The effect of the tides is felt fairly high upstream and can reach the first rapids, sometime as much as 50 kilometers inland (Maripa Falls on the Oyapock, Mapaou Falls on the Approuague, Little Falls on the Sinnamary, etc.).

The mouths of some rivers clearly deviate to the northwest due to the coastal flow (the mouths of the Kaw, Sinnamary, Mana, and Suriname Rivers are excellent examples). Some changes to the shore have provoked the abandonment of coastal villages by indigenous inhabitants, such as Grosse Roche on the Iracoubo. The Amazonian fluvial discharge into the equatorial Atlantic affects the morphology of the shore considerably over the short term, with approximately 280 million cubic meters of fine sediment circulating along the coast of Guiana each year. The shore alternates spatially and temporally between zones of accumulation (consisting of large bands of mud) and zones of erosion (covered by retreating mangrove forests, or sand ridges) (Prost 1992). These rapid, short-term transformations of the shoreline often introduce a significant degree of uncertainty for the establishment of coastal villages. The recent destruction of the ancient Kali'na village at Pointe Isère on the Lower Mana is a key example of this phenomenon, as is the current evolution of the Kali'na village of Awala, where the regression of the shoreline has destroyed buildings as well as a portion of the roadway, and causes problems of salinization for wells close to the sea.

The result of these conditions is that coastal travel is often more rapid by boat (either by river or by sea) than on foot, which explains the excellent navigation skills that characterize the coastal populations. Oyapock Bay and the Île de Cayenne have been interesting stops for maritime voyagers, both in the pre-Columbian and colonial periods, because

> The Montagne d'Argent is located at the bottom of Oyapock Bay, which receives the flow of three large rivers, and offers a secure anchorage for

> several hundred vessels. There, one is perfectly sheltered from the sea by Cape Orange, and it is there that I advise the vessels, which draw more than ten or eleven feet of water, to wait for the high tide to enter Cayenne River (Lescallier 1799:175).

Similarly, in the seventeenth century, "At Cayenne, there is an excellent port where one can load any cargo" (Harcourt [1613] 1926:84) or "I have not found a better port anywhere else along the coast" (Keymis [1596] 1722:265). In addition, the large beaches of Île de Cayenne and the small Bruyère Cove in the Lower Oyapock allow for easy docking of small vessels.

Guiana is made up of four principal landscapes. Moving from the interior toward the coast, these are the *tepuies*, the shrubby savannas, the rainforest, and the coastal plain.

THE *TEPUIES*

The *tepuies* (an indigenous word that means "mountain" or "house of the gods") are the sandstone plateaus with steep and abrupt walls that emerge from the rainforest or the savannas. They are present in the south of Venezuela and of Guyana, as well as along their borders with Colombia and Brazil. They are the result of the uplift of western Guiana that began 180 million years ago. The movement of the terrestrial crust fractured the Guiana Shield, of which many fragments were uplifted to form the *tepuies*. The disintegration of the rocks at the summit of these formations has created caves that are sometimes immense.

There are 115 plateaus of this type known in the region. The most famous is Roraima, at the junction of Brazil, Venezuela, and Guyana, at 2,810 meters asl, with a length of 14 kilometers. Cerro Marahuaca precedes it by only 22 meters. Although Auyan does not quite reach 1,000 meters asl, it extends over 700 square kilometers and is famous for Angel Falls, the tallest waterfall in the world, with a height of 979 meters. The discovery of this cascade itself was remarkable, when the aviator Jimmy Angel crashed into its summit in 1937; his airplane remained stuck there for 33 years.

The inaccessibility of the Guianese *tepuies* provided protection from all European incursions. Once again, due to their strange shape and their isolation, the *tepuies* have inspired fanciful notions. What more mythical places can there be than these lost worlds, pursued by Walter Raleigh and Professor Challenger at the risk of losing their reputations or their heads! It was not so unreasonable for Sir Arthur Conan Doyle to imagine that prehistoric creatures were preserved at the summit of these isolated formations. He was just mistaken about the size of these living antediluvian fossils; his tyrannosaurus and pterodactyls were in reality exceptional endemic microfauna and -flora, preserved from the outside world and evolving in isolation like their cousins on the Galapagos Islands off the coast of Ecuador. Scientists regularly discover new species of carnivorous plants and mosses growing on the wind-battered summits, as well as orchids and unknown bromeliads on the forested flanks of the plateaus. Heretofore unknown insects or frogs are also discovered on these exceptional biotopes. Beginning at about 1,000 meters asl, the summits of the plateaus present a climate that is distinct from that of the rainforest below, colder and more humid, allowing the development of unique species. The *tepuies* thus constitute ecological islands that rise high above the rainforest.

These remarkable plateaus were not, however, inhabited by indigenous peoples; the conditions are too harsh for humans. The Pemon peoples considered the Matawi *tepui* the "place of the dead."

The Savannas of Central Guiana

Enormous shrubby savannas (Figure 18) extend over 54,000 square kilometers in central Guiana from the Roraima of Brazil to the southern border of Guyana and Suriname. These areas have been little explored by Europeans and are still inhabited by indigenous peoples today. In Guyana, they are called "Rupununi"; in Suriname, they are known as "Sipaliwini." While they are relatively flat in the south, they become more irregular to the north, where small hills may emerge (Pires and Prance 1985). They consist of sandy marshes with communities of Moriche palms (*Mauritia flexuosa,* Arecaceae). Although situated in the Amazon Basin, the interior savannas present a distinct climate and flora, with relatively low rainfall that limits the growth of large trees (Daly and Mitchell 2000). The vegetation of these poor-quality, sandy soils generally consists of large expanses of grasses scattered with spiny, stunted shrubs.

Today, humans do not frequent these savannas, but this was not always the case. When the first humans arrived, a semiarid climate, which was much more widespread throughout the Amazon, was dominant (Krook 1979). Thus, the majority of the Paleoindian sites are found in the savannas of central Guiana. These sites consist principally of areas of stone knapping, but evidence of temporary settlement in rockshelters is found in the hills as well (Frans Bubberman, personal communication, 2005).

The Forest: "A Green Paradise"

Dense evergreen forest dominates the Guianese landscape. It covers 90 percent of French Guiana, 80 percent of Suriname, and 75 percent of Guyana. The interior is principally composed of a multitude of small

Figure 18. Sipaliwini savanna, Suriname (photo by Frans Bubberman).

hills in a scattered pattern that forms a relief known as "half-oranges" (outcrops formed of crystalline rocks, eruptive and metamorphic, from the Guiana Shield) (Atlas de Guyane 1979). Viewed from above, this tropical forest resembles an expanse of broccoli, whose canopy reaches heights of 50 to 60 meters. A dense network of rivers and other waterways overlays this ocean of green. In the southern half of the Guianese forest, a series of granitic inselbergs of varied height emerges (some of which exceed 1,000 meters in elevation). These formations may be either isolated or occur in groups. Thus, the interior peneplain of Guiana is shaped by small reliefs, or by fairly high mountains.

Despite this seeming uniformity, the Amazonian forest is actually the most diverse botanical community in the world, with more than 30,000 known species of vascular plants, of which 5,000 to 10,000 are trees (Silman 2007). Nevertheless, the physiognomy of Guiana (and of the Amazon in general) has varied tremendously over the course of the millennia. Between 7,000 and 4,000 years BP, even before human intervention, the fire-prone vegetation and shrubby savannas were more common than present (Sanaiotti et al. 2002; Soubiès 1979–1980). Multiple ancient wildfires ravaged the forest during periods of extreme drought (Charles-Dominique et al. 1998; Tardy 1998; Tardy et al. 2010). The study of pollen from an açaí palm (*Euterpe oleracea*, Arecaceae) peat bog in French Guiana has revealed 3,000 years of vegetation history (Charles-Dominique et al. 1996; Rosique et al. 2000). A period of wetter conditions dominated from 3,000 to 2,000 years BP, as in several other sectors of the Amazon (Van der Hammen and Hooghiemstra 1996). The climate became drier and the forest shrank between 1,700 and 1,200 years BP. Later, the forest resembled more closely that of the modern day, but new perturbations between 900 and 600 years BP provoked the return of pioneer plants. These disturbances diminished progressively between 600 and 300 years BP, allowing the forest to develop its current conditions.

Even until quite recently, it was thought that "these forests have only been weakly impacted by humans, they have also been nibbled on their edges and along their rivers for the exploitation of wood, but even more for the needs of shifting slash-and-burn agriculture practiced by indigenous groups" (Giacottino 1984:19). However, now we understand that, in fact, the impact of pre-Columbian populations significantly shaped the forest in ways that are visible today. At the end of the Pleistocene, humans began to favor the growth of plants that were useful to them (such as palm trees), but the most influential anthropogenic impact began in the Holocene (Balée 1989; Van der Hammen and Hooghiemstra 1996). It is not by chance that 45 percent of the cultivated plants in the Americas come from the Amazon (Clément 1999a, 1999b). In reality, certain landscapes would not exist if humans had not played an essential role in the manipulation, maintenance, and extension of biodiversity (Balée 1989).

Previously, the indigenous groups of the interior did not necessarily always settle close to large rivers (Hurault 1989). It is important to recall that they circulated on foot far more often than they did by boat. Their bark canoes, of simple construction, were not well adapted for the rivers, and even less so for the rapids. In contrast, the network of paths that connected the villages was dense and well maintained. We must imagine the pre-Columbian Guiana as a landscape made up of pathways, upon which existed a constant circulation of hunters, traders, military or ceremonial expeditions, etc. Furthermore, the groups that settled the forest led a seminomadic lifestyle, regularly moving their villages. It is clear that the natural conditions tend to discourage permanent settlements. The exhaustion of the soils, the invasion of weeds, and colonization by leafcutter ants (*Atta cephalotes*, *A. laevigata*, and *A. sexdens*) regularly force communities to relocate after only a short time (McKey and Beckeman 1993). Cultural reasons also explain this high degree of mobility, in particular the loosely developed concepts of territory and ancestor cults. "The geographic setting may encourage or favor, but it never commands absolutely" (Hurault 1989:14).

The forest villages were located in three principal biotopes: (1) the non–seasonally flooded terraces at the foot of hills and bordering small streams; (2) the summit of the hills (dominating the view), with villagers obtaining water from the valley floor below; and (3) the banks of large rivers (this latter was apparently the preferred location for many groups) (Rostain 1994a). The hills present advantages, and by their strategic position, are well adapted to populations living in conditions of conflict. They are sometimes the only possibility for settlement in the Lower Oyapock, where sites were mostly located on the southern side of coastal hills opposite from the sea and close to a stream. Inland, settlement along fluvial banks, particularly near waterfalls, permitted the control of fluvial circulation. Sites are generally located on the high, non–seasonally flooded banks, rarely in a straight line with the river, but rather on the banks of the meanders, and on their points.

A comparison of archaeological and ecological data with ethnographic data helps to understand the decision-making associated with choosing locations for indigenous villages. Several archival texts show that a location's defensive potential was one of the most important criteria considered. The hills are particularly valued in this regard. In 1613, the indigenous village of Caripo, which was led by Martyn, was located in a strategic position on the eastern slope of Bruyère Mountain, at the mouth of the Oyapock:

> When I had thus installed my troupe in this village, I came out to examine its location, and the possibilities for defense. It is a large, rocky mountain, little accessible due to the vast forests, and for the abrupt rocks, but only in some locations, where there were narrow paths, very steep and easy to defend: We thus lagged there as if in a fort, better than that even, for the port was so close that our boats were anchored under our eyes, at the foot of the hill (Harcourt [1613] 1926:78).

This site was thus visible and partially exposed to potential enemies, but also to traders, and permitted control of the passage of boats, facilitating contact and exchange. On Île de Cayenne, the site of Pascaud had a similar position, dominating the mouth of the Mahury and thus controlling access to the river. Located a dozen kilometers from the sea ("three leagues"), the dispersed settlement of the Arikare (*Racalets*) depended on unusual elevations from the marsh of the coast of Amapá and was only accessible by boat (Biet 1664), which offered undeniable strategic advantages.

The villages were also often camouflaged: "the paths that lead to the habitations are so small, that only two men abreast can pass there. The Savages do not want them to be wider, for fear that their enemies will discover them, and one can only reach them by passing through many detours" (Biet 1664:166–7). At this time, a hostile climate was present, notably because of the war between the Kali'na of the coast to the west of the Approuague and the Palikur of the Lower Oyapock and the north of Amapá.

Even the distribution of villages seems to have been a response to the circumstances of insecurity. In 1789, 23 Wayana villages were arranged in a linear group 200 kilometers long, linking the center of Guiana with springs of the Jari, forming a "string of villages spaced along a long line, each spaced a dozen kilometers apart" (Leblond 1813). This arrangement facilitated communication and the rapid assembly of warriors in the case of an attack. Claude Tony (1842) described in greater detail the military aspect of the society and the settlements of the Wayana when he visited in 1796. Palisades began along the roadway 2 kilometers before reaching the villages. This access itself was elevated and surrounded by a cleared zone more than a kilometer wide. In the central plaza, a lookout tower dominated and four sentinels permanently watched the four roadways that led into the village; armed warriors remained in the large central square, and children were taught to use a bow by their elders two times per day. While this military architecture was symptomatic of a constant threat, provoked by the intensification of indigenous movements in the colonial period (Grenand 1971), this type of sociopolitical organization based on warfare implied a long-standing bellicose tradition, such as that of the Tupinamba of the Brazilian coast.

Almost all of the indigenous villages on the coast visited by European explorers were located at the mouths of rivers. This bridging location seemed to fit into a network of intertribal relations. These sites that faced the sea were stopover points for maritime travelers; they also controlled entry and exit to the inland by river. At the beginning of the sixteenth century, on the Guianese coast, "the majority of people are settled close to rivers and travel from place to place by Pirogue" (Harcourt [1613] 1926:85). These villages represented centers of communication and sources of privileged information on the movement of troops and traders. The chiefs of nations generally resided there, welcoming arrivals, organizing warfare and trading expeditions, and managing the territory.

The sites on the riverbanks also presented strategic advantages because they allowed for excellent surveillance of the fluvial circulation, but they were also more vulnerable to surprise attacks. The map of the lower Approuague by Dessingy, created in 1764, shows two locations of settlements that correspond to different groups (Figure 19). Nonetheless, at this time, the distribution of villages was also influenced by the densities of the French settlements. Despite these colonial circumstances, two strategies seem distinguishable among the ethnic groups. The Akokwa and the Norak settled the banks of the river in locations typically used for dockings (falls) or in straight lines. In contrast, the villages of the Kusari, a group more accustomed to walk than to sail, were located on small creeks that flowed away from rivers, indicating that they avoided participating in riverine life. The location

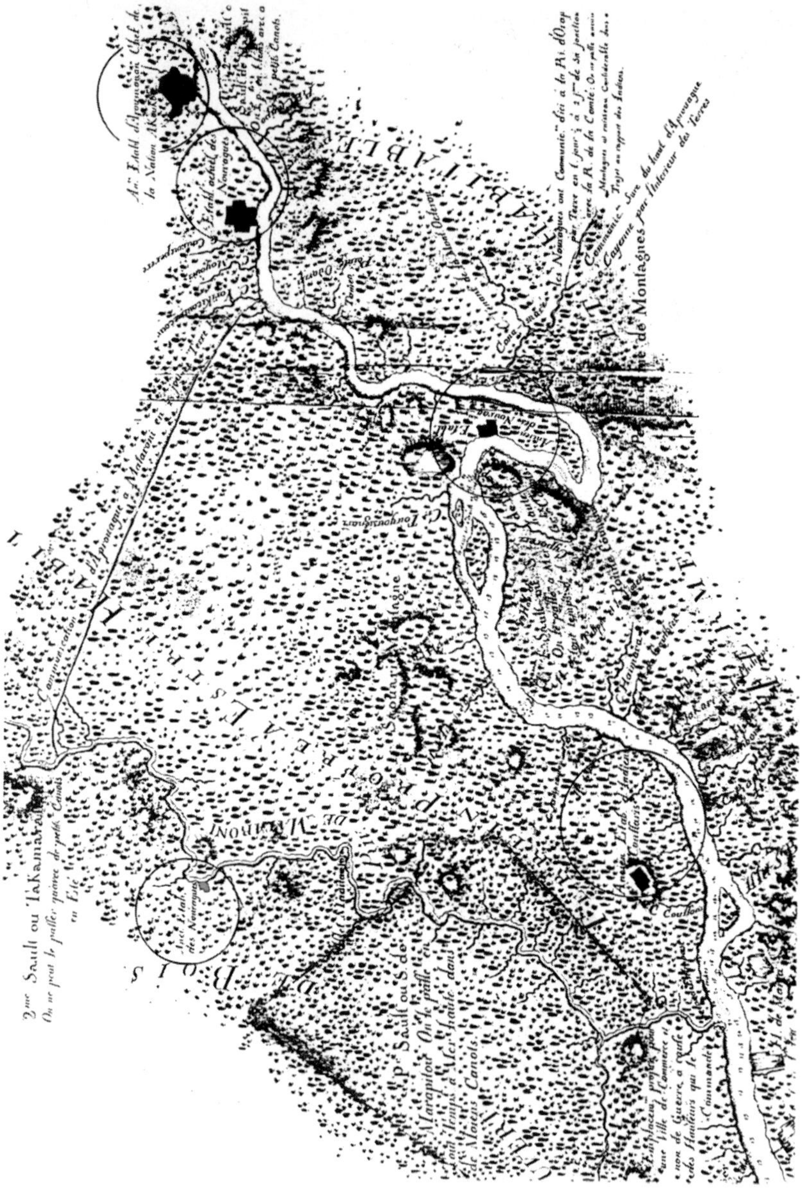

Figure 19. Lower Approuague map of the eighteenth century indicating indigenous villages (with circles drawn by the author) that have been found during archaeological surveys in French Guiana in 1989 (modified from the drawing by Dessingy, 1764).

of villages can thus be an indicator of specific attitudes of different communities.

Today, in the interior, the Wayampi and particularly the modern Wayana settle most often along rivers, but other factors also influence their choices (Grenand 1971; Hurault 1989; Rivière 1984; Schœpf 1972):

- The association of clear water and abundant fish: the falls are privileged because they are usually good locations for fishing. The current renders these places healthy, the rocks permit easy docking, and they do not attract venomous rays (*Paratrygon* sp.).
- The proximity of cultivatable land: the ancient non–seasonally flooded alluvial terraces, not invaded by leafcutter ants, are highly sought after. Clearing for cultivation is generally undertaken a 15-minutes walk from the village.
- The proximity of firewood, with certain species preferred over others.
- Deliberate distance from deep waters, frequented by evil spirits.
- Deliberate distance from ancient villages that were abandoned following one or more deaths.

Data from archaeological excavation, combined with findings from the earth sciences and ethnographic documents, show that decision-making in establishing human settlements responded to very precise criteria. These could be endogenous to the group (cultural, demographic, or technological constraints), or exogenous, such as intertribal relationships and specific environments. This network of relations demonstrates the complexity of factors involved in the location of indigenous settlements. Far from simply submitting to the challenges of the natural environment, indigenous peoples selected the places best adapted to their needs, indicating a profound understanding of their surroundings.

The Coast

The coast of Guiana is part of the vast Quaternary sedimentary plain that extends more than 1,600 kilometers from the mouth of the Amazon to that of the Orinoco. It measures 90 kilometers wide in the Nickerie district in Suriname. The coastal plain then narrows when it reaches French Guiana, only spanning 5 kilometers to the west of Île de Cayenne, and on average 16 kilometers farther to the east.

It is only in French Guiana that outcrops of the shield occur on the shore, forming small elevations. Remarkable high spots on the coast are located at Île de Cayenne, where they can measure up to 162 meters high (Mont du Montabo, Mont Bourda, Montravel, Montagne de Rémire,

Montagne de Mahury, Mont Cabassou, Montagne de Tigre, and Mont Saint-Martin) (Figure 20). They are also found in the Lower Oyapock, where the highest massif is 341 meters tall (Monts de l'Observatoire, Montagne Bruyère, Massif des Trois Pitons, and Montagne d'Argent). A few hills, up to 152 meters high, appear at the mouth of the Kourou River (Montagne des Pères, Montagne Lombard, Montagne Café, Monts Pariacabo, and Montagne Carapa).

At the time of European contact, these hills were good landmarks for navigators: "We saw two high Mountains (hills of Ouanary). Several Rivers flow into the Sea along the Northern and Northwestern Coast. We anchored close to these mountains" (Keymis [1596] 1722:262). These are the only obvious high points visible and accessible from the sea between the Amazon and the Orinoco, which explains their interest for the first explorers of the coast.

Since the seventeenth century, colonists contrasted *terres noyées* (flooded lands) or *terres basses* (lowlands)—this term was still being used by Samuel Guisan in his 1788 treatise on polderization—with the *terres hautes* (highlands). The first category refers to soils of marine origin and includes the mangroves and marshes (*savanes tremblantes* or *noyées*). The second category includes the sand bars and coastal hills that were long preferred for settlement and agriculture by the Europeans.

In any case, seen from the sky and from the sea, the landscape is one of low coasts that are often marshy, bordered on the Atlantic coast by mangroves and by swamps dominated by *Euterpe oleracea* palms, and

Figure 20. Bruyère Hill in the Oyapock Bay, French Guiana, is one of the few coastal elevations between the Amazon and the Orinoco River (photo by Stéphen Rostain).

bordered to the south by the northern extremity of the inland rainforest. This coastal plain covers 230,000 hectares between Île de Cayenne and the Maroni River. The mangrove swamps are distinct from the marshes. The former are brackish-water environments that are periodically submerged by the tide and covered by mangroves, whereas the latter, farther inland from the coast, are fed by groundwater and rain. The coastal band combines saltwater, brackish-water, and freshwater environments, mixing flooded and nonflooded land, as well as different types of sandbar and dune habitats and a relatively diverse flora.

This coast was long a zone sought after for settlement by certain pre-Columbian populations. They could access relatively easily resources from the forest, the swamp, and the ocean, as well as saltwater, brackish-water, and freshwater zones.

A distinction exists between the recent and the ancient coastal plains, which have very different characteristics and offer different conditions for settlement.

The Recent Coastal Plain

The recent coastal plain is low and marshy, with elevations below 4 meters asl. It is bordered along the maritime side by mangrove forests, which generally are located a few dozen centimeters above the highest tide level; these forests are particularly subject to submersion by the tide. The mangrove forests are flooded by seawater up to 2 kilometers inland during the tides associated with the new and full moon.

Three phases of accumulation have been identified in the Demarara Formation (Turenne 1979): Mara, the oldest (8,000–6,000 years BP); Moleson (2,500–1,300 years BP); and Comowine, the most recent (beginning at 1,000 years BP). Fine accumulation of marine and fluviomarine facies characterizes the Mara Holocene transgression, as well as phenomena of progradation, such as those that are currently produced along the shore. The vegetation is essentially composed of white mangroves (*Avicennia germinans*, Avicenniaceae) and brackish marshes dominated by *Eleocharis mutata* (Cyperaceae), *savanes à palétuviers morts*, and lagunas.

The deposits include fine clay and clayey-silt sediments that originate from the Amazon. They also include coarse and medium-grain sands that make up the sand ridges that have formed parallel to the shore and are separated by marshes. The environment of the low plain is therefore a mixed wetland, made up of seasonally flooded zones and marshes (in the middle of which appear the sand ridges), as well as mounds with scalloped edges, called "islands of dissection" (Prost 1992). The sand ridges result from ancient beaches that formed under erosive conditions and in a high-energy environment. They measure several dozen meters wide and dozens of kilometers long. These formations began to form directly on the ancient coastal plain. Although the younger coastal plain is not very wide in French Guiana, it is, in contrast, much broader in Suriname and Guyana, where, due to the richness of the sand and shell ridges and the fertile silts, it was densely occupied prehistorically, and was the focus of the most significant pre-Columbian cultural developments.

Nonetheless, the form of the coast of Guiana changes frequently and rapidly, and its current conditions are very different than those of the pre-Columbian period. The shoreline responds very quickly to changes in the phases of accumulation and erosion. Banks of silt are deposited by discharge from the Amazon, and stretch along the coast. These are interspersed with gaps that generally have formed sandy beaches (Prost 1987a, 1987b, 1990a, 1990b). Although the area west of Kourou presents an attractive beach today, it was covered by an enormous deposit of mud and mangrove forest only 25 years ago. This dynamic coastal environment has frequently led to the destruction of coastal settlements. Moreover, sites that were originally established on the shore are now found much further inland due to movement of the current coastal plain and, conversely, some sites that were more distant from the sea at the time of their occupation now are located closer to the shore.

The study of sites on the coastal plain depends on a clear understanding of the phenomena that have transformed (and continue to transform) the shoreline, and more generally by the geomorphological history of the coast.

The recent Kali'na site at Pointe Isère at the mouth of the Mana River, in French Guiana, is an interesting example of coastal dynamics and of the destruction of coastal sites. Study of historic maps shows that Pointe Isère began to form around 1865 and that it has been subject to strong erosion in the last 50 years. A Kali'na village was formerly ideally located on a sandbar facing the ocean. At the end of the 1940s, the accumulation of silt and the proliferation of mangroves in front of the hamlet forced the Kali'na to give up this location (Delawarde 1967). A phase of shoreline regression followed this period of accumulation, and the village, attacked by the sea, was definitively abandoned around 1950 (Abonnenc et al. 1956). The last traces of the village were finally destroyed in 1990 (Cornette et al. 1992). The ocean continued to destroy the peninsula of Pointe Isère over the years. Finally, in 2006, the constant erosion wore through the remaining tongue of sand 5 kilometers from its end (Figure 21). The opening progressively continued to grow, and today it measures 2 kilometers wide where the Mana empties into the ocean. At the same time, a sand bank was deposited at the former mouth of the river, sealing it off completely. Then, a sand ridge formed, closing off the estuary. The alluviation from the river gradually accumulated behind the sandbar, and without a direct evacuation toward the ocean (since the river's mouth is now farther upstream), a marsh is currently in the process

Figure 21. Natural opening of the right bank creating a new mouth to the Mana River, French Guiana, simultaneous to the formation of a sand ridge (*chenier*) at the previous mouth (Image © 2012 IGN-France. Image, U.S. Geological Survey).

of forming. It is easy to envision future phases of this geomorphological evolution. The former estuary will continue to experience sedimentation until it is completely infilled, creating a wide marsh that will be gradually colonized by sedges and other aquatic plants (Figure 22). A flooded zone will then form in front of the village of Awala, much to the dismay of its inhabitants. Similarly, the new mouth of the river will widen under the force of the current, while Pointe Isère will be transformed into a small island on the left bank and in front of the new swamp. In the long term, a wide mouth will form to the north with a marshy left bank.

The establishment of a village on the shore depends on geomorphological conditions. Phenomena of shoreline regression, often rapid, can force inhabitants to abandon a location. These processes also lead to the disappearance of sites. The embankment on which the site of Rorota was constructed on Île de Cayenne was strongly attacked by the sea during the early 1980s (Rostain 1994a). In contrast, phases of shoreline accretion have probably pushed formerly coastal sites into the interior. Former shoreline sites today may have disappeared under the ocean, or are now found in the interior of the lowlands.

The system of Amazonian fluvial discharge into the equatorial Atlantic modifies the shore's morphology significantly and rapidly. Approximately 280 million cubic meters of fine Amazonian sediments circulate along the coast of Guiana each year. The shore alternates

Figure 22. Mouth of the Mana River closed in 2008 by a *chenier*. The Kali'na village of Awala is visible to the left, and the Piliwa raised-field complex is in the foreground (photo by Stéphen Rostain).

spatially and temporally between zones of accretion, occupied by large deposits of mud, and zones of erosion, represented either by mangrove forests or sand ridges (Prost 1992). Clearly, the morphological transformations of the shoreline in the short term render the establishment of villages quite precarious.

The swampy coast populated by mangroves offers few possibilities for human settlement. One of the indigenous responses to this mixed wetland environment was to raise their settlements on stilts (Figure 23); the village of Maraon, composed of three large houses on stilts, was observed in 1623 by Jesse de Forest on the shore of the Île de Curuá (*Sapno*) at the mouth of the Amazon. Whereas in settlements built directly on the ground trash tends to be deposited behind familial residences or slightly farther away, in the case of houses on stilts, refuse can be disposed of directly underneath the house: "The floor they make out of hurdle is used to let drop everything through and they . . . even urinate and defecate without going out" on the County (*Oyac*) in 1688 (manuscript Goupy-des-Marets 1690). Finally, there are the incredible houses on stilts of the Maye and the Karanariou of the Amapá coast, which use living mangrove roots as supports (Chrétien 1719).

Indigenous people have developed other solutions to inhabit flooded zones. The fisher-gatherers of the Alaka culture, in western Guyana, raised the floors of their settlements to live close to the shore. They piled

Figure 23. Indigenous pile dwelling on the swampy coast of Suriname (from Dubois and Guy 1897:234).

the shells of marine resources they consumed, gradually forming enormous mounds above the flood level. In western Suriname and eastern Guyana, to occupy the swampy lowlands of the recent coastal plain, the populations of Barrancoid and Arauquinoid traditions also built settlement mounds and traveled along causeways or used canoes to cross the swamp. The characteristics of the coast strongly influenced the models of indigenous settlements because the absence of sand ridges in this region forced the inhabitants to build artificial mounds, using bricks made of clay from the swamp.

Beyond their function to raise structures well above the flood level, the mounds also may have served specific cultural and social functions. On Marajó Island, where the mounds built by the Marajoara culture are much higher than the current high-water level, Roosevelt (1991) suggested they served a defensive function. However, variations in sea level over the Holocene may better explain the height of the mounds at the time of their occupation.

Small beaches protected by rocky outcrops were also occupied. Sedimentological, mineralogical, and chemical analysis of the site of Rorota, built on a sandy embankment in French Guiana, has shown that the archaeological level rests on a layer representing a regression of the sea, intercalated between two periods of transgression. Stratigraphic study of these deposits demonstrates that the archaeological deposit is dated to the end of the Demarara phase, around 1,000 years BP (Seurin 1976). The site clearly precedes the second half of the seventeenth century because one of the first French settlements of the *Seigneurs de la*

Compagnie de la France Equinoxiale was established here in 1652. Moreover, the French encountered a Kali'na village on the shore, farther to the east, at Pointe Mahury: "the other Cabin where lived the Savage Apoto, Son-in-law of Bimon, was just at Pointe Mahury, about two hundred paces from that of Bimon" (Biet 1664:162).

Nevertheless, the first European voyagers seemed to have only come into contact with coastal villages at the mouths of rivers or on islands.

The Ancient Coastal Plain

The ancient coastal plain is located behind the recent coastal plain and maintained by the forest of the interior. It presents a landscape quite distinct from the formations that surround it to the north and the south. Situated between 4 and 15 meters asl, it is a landscape of savannas (high, low, or forested), small forested is lets, and riparian forests.

It appears that its soils are potentially among some of the least poor of the lowlands. With control of water, it is possible to gain from this slight advantage. Although the chemical and physical constraints of the soil are significant, cultivating with raised fields permits more effective regeneration of the soils than with slash-and-burn agriculture in the highlands, and thus permits longer-term use.

The irregular landscape of the coastal plain is characterized by wide sand-silt ridges, called sandbars, which have developed over a substrate of marine clay that formed from the Pleistocene Coswine Formation. The sandbars present differences, particularly with regard to the type of crest and their orientation, due to their morphological and pedological evolution. Sometimes these sandbars are long and separated by small, narrow and hydromorphic basins; in other cases, they are simply sparse assemblages of varied shapes that occur in the middle of flat zones and flooded depressions. In whatever case, one thing is certain: due to their size and location in the middle of the seasonally flooded savannas, the sandbars have always served as prime locations for human settlement.

The sand ridges are a different formation than sandbars because they correspond to the remnants of former shores. Those of the Holocene plain are unusual in that they are formed by deposits of local medium- and coarse-grained quartz sand (Demarara sand) and always occur over a substrate of fine sediments that are composed of Demarara clays. They were formed between 5,000 and 3,000 years BP and derive from ancient cordons (Barone-Visigalli et al. 1991). The ancient coastal plain of Guiana, made up of Quaternary sediments, thus presents two facies of former sandy shorelines, the sandbar and the sandy ridge, upon which many pre-Columbian sites were built. Although these two landforms correspond to different sandy coastal ridges, their geologic, pedologic, and stratigraphic characteristics are different. In contrast, the archaeological traits of the sites that were built upon these elevated locations are almost identical. These similarities do not, however, signify that the ancient occupants did not also consider the environmental properties of each formation when constructing their settlements.

The Sites of the Île de Cayenne Sandbars

The Pleistocene sandbar, with a sand-clay texture, is the result of the transformation of a covering that was initially ferralitic, but which then diminished in clay and iron, resulting in the formation of a podzol at the summit, and a hydromorphic leached soil at the bottom of the sandbar (Grimaldi et al. 1992). The sandbars, bordered along their axes by marshy drainages, and which extend across the ancient coastal plain, measure 500 to 1,000 meters long, 400 to 600 meters wide, and 4 to 15 meters high (Grimaldi 1990). They are composed of Coswine sands, fine and sorted, slightly clayey (20 percent clay), with some grains of mica. They become coarser towards the base, where they mix with Continental series. The sandbars rest either on marine clays subdjacent to the Coropina series, which are white with red striations, compact, and formed by slightly permeable clays with low rates of salt, and are composed of 60 percent kaolinite and 40 percent silts, or they more rarely rest on the alterites (Turenne 1979). The free vertical drainage allows for the evacuation of water during the rainy season, resulting in a hygienic living biotope. The sandbars are also among the best locations for human settlement.

Sites on the Île de Cayenne sandbars present similar characteristics with regard to their location, size, stratigraphy, and the archaeological material they contain. Although sherds are always visible on their surfaces, the archaeological deposit, measuring between 10 and 25 centimeters thick, generally appears between 30 and 50 centimeters below surface. Sherds are sometimes present at a depth of almost 150 centimeters, but the archaeological deposit generally is not more than 25 centimeters thick. It was less than 20 centimeters at the site of Thémire, home of the eponymous culture, because the surface had been razed for construction. At this site, floors covered with sherds forming layers approximately 10 centimeters thick were recorded 20 years ago (Rostain 1994a). They are likely the remains of areas of refuse disposal that surround pre-Columbian houses. A very large archaeological site is found on the series of coastal ridges bordering Mont Saint-Martin to the west and Montagne de Rémire to the east, where various archaeological excavations have been carried out over 25 years. Despite the different names given to each of the excavated zones (Les Âmes Claires, Vieux Chemin, Mini-Circuit Automobile, Glycérias, and Montjoly Bar) they very likely all constitute a single large settlement along the sand ridge. The similarity of the remains and their proximity strongly suggest a single occupation of the site. The site likely extended over 1,500 by

400 meters (or 60 hectares), which is not extraordinary when compared with modern indigenous villages on the coast. For instance, the Kali'na settlement of Awala stretches over 2 kilometers of beach at the mouth of the Mana, with a width of approximately 500 meters.

The Sand Ridges of the Western Part of the Coast of Guiana

The coastal plain of Guiana contains long and narrow sand ridges, known as *cheniers* in French, arranged approximately parallel to the shore (Figure 24). "A *chenier* is a type of sandy beach (or one very rich in shell fragments) formed on a substrate of fine sediments (clays, silts, etc.) on a coast characterized by phenomena of accretion and regression" (Prost 1990c:19). These *chenier* plains occur throughout the world, from the equatorial zone (such as in Africa) to the higher, glacial latitudes (such as in Alaska). Furthermore, these formations have often served as favored locations for human settlement since ancient times. In Guiana, these ancient beaches, arranged in bundles and isolated from the shore by a marshy space, make up a plain of *cheniers* open to the ocean, the evolution of which is driven by the dynamics of the shore. The *cheniers* measure 10 to 200 meters wide and 2 to 4 meters high (Tacoma et al. 1991; Turenne in Atlas de Guyane 1979). They are made up of local, medium- to coarse-grained quartz sands (formations of altered basement rock of the Guiana shield, Detritic Base, and Coswine series) transported by rivers and transformed by marine action. They may also contain sediments from the Amazon (Krook 1979). Two principal systems of *cheniers* are known in Guiana (Prost 1990a): the "ancient" ridges that intercalate with the marine clays of the Mara phase (8,000–6,000 years BP) and the Moleson phase (2,600–1,300 years BP), and the "recent" ridges that are situated between the Moleson and Comowine phases (1,000 years BP to the present). Currently, these *cheniers* continue to form in the spaces between banks, when the deposition of fine Amazonian sediment is interrupted (at the Île de Cayenne and Kourou, for example), and also in zones affected by erosion (such as Pointe Isère in the Lower Mana).

Numerous archaeological sites have been observed on the coastal *cheniers* of the Île de Cayenne at the Orinoco Delta. The pre-Columbian occupations typically belong to the Arauquinoid tradition, but survey has also demonstrated a strong presence of Kali'na, the dominant ethnic group on the coast since European contact. Moreover, numerous ancient maps indicate indigenous villages along the coast, probably established on these formations.

The sandy Quaternary ridges of the western part of the coast were densely occupied due to the numerous advantages they offered: land that overlooks the surrounding marshes, soil above the flood level all year round, and easy and rapid access to the terrestrial and marine resources. From 2,000 to 1,300 years BP, a series of *cheniers* developed along the

Figure 24. *Cheniers* of the western coast of French Guiana (photo by Stéphen Rostain).

coast of Suriname (Versteeg 1985) and probably along the entire coast of Guiana. The site of Kwatta-Tingiholo in central Suriname was constructed on a *chenier* whose formation has been dated to approximately 1,715 years BP (Roeleveld and van Loon 1979). It was occupied roughly between 1,510 and 970 years BP during a period when the *chenier* faced the ocean. The possibility cannot be ruled out that certain *chenier* sites did not border the Atlantic when they were occupied.

The thinness of the archaeological level encountered in the settlement sites, between 10 and 40 centimeters below surface, indicates a single occupation, and the abundance and stylistic homogeneity of the remains suggests it was dense.

These locations continued to be preferred by indigenous peoples after the Conquest. During his voyage on the Comté in 1688, Groupy-des-Marets spent the night in an indigenous village of 30 to 40 inhabitants, 60 kilometers ("15 leagues") from Rémire: "first an Indian house is always built either in the flooded savannas or on the bank or close to a river, a creek or a lake, such that its construction is particular" (manuscript Goupy-des-Marets 1690). After the Conquest, the Kali'na sometimes settled on hilltops, or more frequently, on the sandy coastal ridges and on the shore of the ocean: "to the West of the island (of Cayenne), there is a long field of two leagues seeded with Caribe villages. . . . Manamanori, Icarouari, and Synamari are among the rivers inhabited by the Caribes" (Forest 1625:52). This settlement corresponds to that of the Arauquinoid archaeological sites that extend along the coast. These villages, not visible from the sea and along a shore with a difficult anchorage, are less accessible from the exterior than those at the mouths of rivers. Despite their more remote positions, these sites remained linked to the ocean. Northwest of Kourou, artificial levees cross the low marshy zones and join two coastal ridges parallel to the shore, which likely allowed for direct access from the villages to the sea. These paths may have followed the ridges, as they do today: "It is through these Savannas that one can go by land from Kourou to Suriname . . . [a] route which is very familiar to the Savages from this area" (Barrère 1743:27). This linear distribution facilitated contacts among villages, perhaps reflecting an attempt to reduce conflict. It seems that at that time of European contact the population of the coast of Guiana had increased considerably and the growth of villages had led to an intensification of territorial warfare. The numerous intertribal conflicts of the colonial period, such as those between the Kali'na and the Palikur of French Guiana, often resulted from the forced migrations that were provoked by the Spanish and Portuguese hunt for slaves. The conflict may also have reflected the lingering influence of warfare that existed before the arrival of the Europeans. Nonetheless, a need for protection was not necessarily always a primary concern, and a variety of other factors (including economic and social factors) may have determined the location of villages, as they do today. Some things taken into account in the planning of contemporary indigenous villages include the proximity of a source of fresh water, abundance of game, zones appropriate for farming and the collection of firewood and construction wood, the condition of the soil in the rainy season, and, sometimes, a certain distance from cemeteries.

The modern Kali'na villages are still located on *cheniers*, on the seashore (Awala and Yalimapo in French Guiana), or more inland (Bellevue in French Guiana), on the banks of large rivers (Christiaankondre and Langamankondre in Suriname), or on the edge of the savanna (Coswine, Île Portal in French Guiana).

Finally, the seasonally flooded savannas and the marshes to the west of Île de Cayenne are frequently covered with hundreds of mounds, which are absent from savannas further east, like the savannas of Amapá. What are these structures? It is first necessary to determine if they were natural or human-made. A number of regular and symmetrical structures that are nonetheless completely natural in origin are known around the world, such as gilgaïs (small mounds) or *touradons* (Rietkerk et al. 2004; Rietkerk and van de Koppel 2008). Such natural features have been observed in other seasonally flooded savannas of South America. In temperate and tropical regions, gilgaïs may form under conditions of contrasting rainfall regimes because of the presence of fine textured vertisols with an abundance of swelling clays (montmorillonites) (Buol et al. 2003). "It consists of a succession of closed microbasins and micromounds on an almost horizontal surface or microvalleys and microcrests, between a few centimeters to 2 m high, which follows along a slope" (Lozet and Mathieu 1986:94). Similarly, a great number of regularly spaced mounds that were constructed by termites can be found on seasonally flooded areas of the Llanos of Bolivia, Colombia, and Venezuela (Mayle et al. 2007). Thus, archaeological data are frequently vital to distinguish between landscapes that are completely natural in origin, and those that are anthropogenic.

Nevertheless, the mounds of Guiana are characterized by regular geometric patterns that are undoubtedly human in origin. Furthermore, the long, rectilinear mounds, which are sometimes associated with round mounds, do not exist naturally. The locations of the mounds relative to the topography of the sites also indicate a structured organizational scheme. The large raised fields are located in the most flooded zones, and the medium raised fields follow the curve of thalwegs in the savannas. Most raised fields are located along sandbars; small and medium raised fields cover the seasonally flooded savannas that are dry in August. This highly systematic spatial organization indicates human intervention and reflects a differential management of areas with respect to their drainage. The organization of the mounds as a function of drainage constraints is also observed within certain sites themselves. Moreover, the mound complexes are always located to the south of *cheniers* where pre-Columbian villages were present.

These different observations demonstrate that, at least for the savannas that have been most intensively studied, humans principally created the marked topographic heterogeneity seen today (McKey et al. 2010). While shifting agriculture currently dominates in the Amazon, diverse and elaborate agricultural techniques were developed during the pre-Columbian period, such as the farming of *terra preta*, farming the humid floodplain during the dry season, or the use of raised fields on the seasonally flooded savannas. This last technique was amply used in various zones of the Amazon with a strong contrast between seasons in the last 2,000 years.

Contrary to the beliefs of European colonists, the coast offered a large range of edible resources. First, there was a rich and diverse fauna, which were easily hunted. More than 200 species of birds live along the coast of Guiana: ducks, herons, egrets, frigatebirds, terns, red ibis, etc. (Hansen and Le Dreff 2002). From August to October, numerous species pass along the coast of Guiana during their southward migration. In January and February, as well as April, two waves of birds flying north from Brazil cross the coast. During these periods, thousands of birds cover the marshes and coastal islands of Guiana:

> The birds that live along the sea in the marsh are very many & very good to eat. . . . I would say that what seems incredible and which is nevertheless true, is that there is such an abundance of all sorts of birds, that there are islands that are full of them. Some of the Islets seem to be on fire, when they are covered by the birds that are called *Flamans* (Biet 1664:344–5).

Even if contemporary hunters have ravaged the area, there was previously an abundance of game on the coast: deer, tapirs, peccaries, armadillos, caimans, iguanas, and agoutis. The ocean and the rivers were also very rich in fish, allowing the indigenous people to develop very diverse and effective methods of fishing (Hurault 1989).

Large marine vertebrates, much more numerous in the past, frequented the coast. The manatee, or "sea cow" (*Trichechus manatus*), weighs on average 500 kilograms, but can reach as much as two tons. It is found along the coast and rivers as far as 100 kilometers upstream. The population of this herbivore decreased in the colonial period, but it was very common before European contact and frequently hunted by indigenous groups. More than half the total population of leatherback sea turtles (*Dermochelys coriacea*) visits the coast of Guiana each year. This species is one of the largest, 2 meters long and weighing between 250 and 1,000 kilograms. Marine turtles come between April and August to lay their eggs on the Guianese beaches. They dig holes in the sand and then deposit more than 100 eggs at a time. A female can lay up to seven batches of eggs per season. At the end of the 1980s, more than 40,000 nests were counted annually on the beaches of Suriname and French Guiana. Thus, as many as 2,000 turtles can come to lay their eggs in a single night (Fretey et al. 2002). Indigenous groups traditionally consumed the eggs of marine turtles in large quantities. Certain clues indicate that the turtles were also caught to be eaten later. A Dutch map from the eighteenth century indicated three turtle pens (*Parken voor Schildpadden*) at the mouth of the Iracoubo River, to the west of French Guiana. This animal was also frequently represented in the ceramic art of coastal pre-Columbian populations, which indicates its importance in their imagery.

This brief overview shows that coastal populations had direct access to a wide range of marine, coastal, and inland animal resources. Furthermore, the indigenous technique of smoking allowed the building of stores.

A Virgin Forest?

In the Amazon, the omnipresence of vegetation is generally considered an obstacle to the discovery and excavation of archaeological sites. Strangely, this flora has long been thought to have made the establishment of pre-Columbian settlement difficult; as a result, few reconstructions of the vegetation of pre-Columbian sites have been attempted (Balée 1987, 1989). Nonetheless, certain plants represent traces of ancient human activities, such as the species cultivated, and the secondary vegetation that thrived in slash-and-burn fields and in villages. Between 140 and 200 years after the abandonment of a slash-and-burn field, primary vegetation replaces the secondary vegetation (Uhl and Sadarriaga 1987). Nevertheless, certain associations of species that result from human activities remain apparent, among them ceiba (*Ceiba pentandra*, Malvaceae), guapinol (*Hymenaea courbaril*, Fabaceae), and bamboo (*Guadua* sp., Poaceae), and perhaps mahogany (*Cedrela odorata*, Meliaceae) (Versteeg and Bubberman 1992). Therefore, the traces of ancient settlements in the tropical forest may sometimes be clearly preserved in the modern vegetation (Figure 25).

The species most useful to humans, such as the cashew (*Anacardium occidentale*, Anacardiaceae) in particular, pineapple (*Ananas comosus*, Bromeliaceae), or wildcane (*Gynerium sagittatum*, Poaceae) (Hoock 1971), are common in the high savannas of the coastal plain of Guiana. Have the plants they cultivated left living descendants? Caution is required in addressing this question, because some of them have wild relatives that grow naturally in Guiana. Genetic studies of *Manihot* growing in coastal savannas, for example, showed that its populations are those of a close wild relative of manioc, bearing signatures of only very local and recent instances of introgression from domesticated manioc (Duputie et al. 2007; Léotard et al. 2009).

On the other hand, archaeobotanical studies of vegetation that originated from anthropogenic activities have provided interesting

Figure 25. "Through the bamboos." Such vegetal formation often results from a previous human settlement (Crevaux 1882:249).

data in Suriname, where numerous archaeological sites in the interior have been discovered through observation of the current surface vegetation (Frans Bubberman, personal communication, 2005). The plant communities present on closed anthropogenic sites were studied on four artificial settlement mounds of the western coast (Buckleburg-1, Wageningen-1, Hertenrits, and Burnside) (Werkhoven and Versteeg 1980). A specific combination of plants was observed, different from those found on the sand and shell ridges of the coast. In Suriname, the density of economic species used by modern indigenous groups is generally from 10 to 15 percent in Suriname (note that construction wood is not included in this calculation). It rises to 47 percent on the pre-Columbian settlement mounds, which is much higher than in settings that are environmentally similar (such as the coastal *cheniers*), but have not been recently subject to human impacts (Werkhoven and Versteeg 1980). This study, which yielded interesting data, merits further investigation, particularly to increase the inventories of plants that occur in natural versus anthropogenic biotopes. However, it must be emphasized that the plants mentioned included several wild relatives of domesticated plants (manioc, sweet potato, pineapple, etc.) that frequently grow in disturbed sites whether the disturbance is of human or natural origin. Nevertheless, the high proportion of economic plants leads one to think that the modern vegetation of the mounds is partially a remnant of the indigenous occupation of these locations. The frequent farms close to the houses of modern indigenous villages (achiote, cashew, cotton, peppers, fruits, etc.) seem to support this hypothesis.

It is probable that pre-Columbian populations intentionally encouraged assemblages of species in the Amazonian forest and carried out profound and sustainable environmental transformations. This process began with the first Paleoindian hunter-gatherers who would have burned the savannas to expand them and push back the forest (Van den Hammen and Hooghiemstra 1996). They also very likely developed a deep knowledge of the plants and carried out the first selections for useful species. Human impact exploded with the adoption of agriculture and the later development of large-scale agricultural systems, such as raised fields and the cultivation of *terra preta*. These changes represented much more than a simple adaptation to the environment; they demonstrated a true landscape management in some areas of Amazonia. Some have even proposed that "the human species is itself a principal mechanism of change in the natural world, a mechanism qualitatively as significant as natural selection" (Balée and Erickson 2006:5). Clearly, not all researchers would agree with this statement, particularly because it ignores a significant difference in scale. Humans affect the species composition of plant communities and the environment more than they do the evolution of individual populations, whereas natural selection functions at the genetic scale within populations of each species and has done so for millions of years (Doyle McKey, personal communication, 2010).

Recent advances across several disciplines have greatly advanced our understanding of the historical ecology of Amazonia. The "Archaeology and Ecology of the Coastal Savannas of Guiana" project, carried out in the last few years in French Guiana, may allow us to propose an intermediate position between the supporters of the exclusive action of nature and the defenders of the primacy of human intervention. It is indeed the confluence of these two forces that have produced the modern landscape (McKey et al. 2010). Although the concept of ownership of land may have been absent in the indigenous world, the spaces controlled by Amazonian groups were always carefully organized and managed. Thus, although the alliance between Guianese nature and indigenous peoples was generally useful for mankind, it unfortunately often degenerated under European influence.

Chapter 3

Terra Cognita: 10,000 Years of Human Impact

In the last 20 years, Amazonian archaeology has improved our understanding of the history of the peopling of the region. Some authors have proposed chronological reconstructions of the pre-Columbian Amazon (McEwan et al. 2001; Neves 2006). As useful as these studies have proven, they are often focused on the Amazon Basin and ignore Guiana. As a result, the coastal Amazon is generally absent from chronological schemas.

Several authors have published chronologies for Guyana (Plew 2005; Williams 2003), Suriname (Versteeg 2003), and French Guiana (Rostain 1994a). However, they are limited by the political boundaries of modern states, which does not allow for a global perspective of ancient settlement. Several years ago, I proposed a chronological schema for the entire region of Guiana (Rostain 2003, 2007, 2008c). Although this framework is still generally valid, more recent discoveries in the various countries that make up Guiana call for revisions to certain aspects of the chronology. In addition, my recent studies of artifact collections from Guiana Island have given me a new perspective on the region's material culture. In any case, the history of the pre-Columbian peopling of Guiana Island cannot be separated from that of the Amazon as a whole. Several conceptual bridges between the two regions are proposed below.

Humans have been present in Guiana for almost the last 10,000 years. Originating from Central America to the northwest, they initially colonized the large central savannas and the Middle Amazon before progressively settling the rainforest along the course of rivers and the coastline. It has long been inferred that the Amazon existed in the shadow of the Andes as a secondary cultural development; the lowlands themselves were not a center of innovation. Yet, in the last 20 years, we have begun to realize that the Amazon has been the birthplace of important human developments, such as ceramics and the domestication of plants. More recently it was settled by a variety of large societies, perhaps stratified. Nevertheless, while certain regional developments occurred with Guiana itself, the major peopling of the Common Era diffused from the cultural hearths of the Middle Orinoco to the north and from the Lower Amazon to the south.

I differentiate between five principal cultural stages in the history of pre-Columbian settlement. I propose to modify slightly and to tropicalize an English children's nursery rhyme, also used by Hercule Poirot, to summarize the successive stages of pre-Columbian settlement in Guiana:

This little tapir went to market
The second little tapir stayed home
The third little tapir ate roast beef
The fourth little tapir had everything
The fifth little tapir cried "ay! ay!"

The first tapir represents Paleoindian hunter-gatherers. The second represents the beginnings of sedentism with the fisher-collectors that inhabited the riverine and maritime shores. The third designates the appearance of agriculture and food production. The fourth refers to the large, complex societies that developed intensive modes of production, such as raised fields and the storing of surplus. The final tapir refers to the Conquest and its disastrous consequences for indigenous people.

Despite the following chronological presentation, it is important to bear in mind that to consider the history of settlement in the Amazon as a linear schema that begins with hunter-gatherers and ends with stratified agricultural societies is an oversimplification. The development of one practice, such as agriculture, does not necessarily replace others, such as hunting, fishing, or gathering. Ceramic production does not always accompany sedentism, and vice versa (Niederberger Betton 1987). The adoption of one or another trait must not be seen as the mark of a strict evolutionary stage, but rather as the result of carefully considered choices made by populations that managed and transformed their tropical environment for thousands of years.

The scale of cultural change in the pre-Columbian Amazon contrasts with that of other regions, such as the Andes or Mesoamerica. The spatial and chronological extent of the cultural traditions is particularly important. Certain developments may have appeared over a large portion of this region (e.g., similar to the size of Europe). The Polychrome tradition extended from its birthplace in the Lower Amazon all the way to the Andean piedmont, 3,000 kilometers away. Similarly, this large diffusion often doubled in the long term. Cultural traditions maintained themselves easily over more than a millennium, some continuing to exist after 2,000 years. The same Polychrome tradition debuted around AD 300 to 400 and developed all along the Amazon Basin until the Conquest. It continues to survive even today through the art of the Kali'na of Guiana, the Shipibo-Conibo of Peru, and the Quechuas of Ecuador.

It is essential to recall the exceptional geographic and temporal hypertrophy of the Amazonian cultural traditions to gauge the pre-Columbian evolution of the region.

The Initial Settlement

The current data suggest that humans arrived in the Americas at least 30,000 years ago (Lavallée 2005). They came from the northeast of Asia, crossing Beringia, which at that time was close to 1,000 kilometers wide. After crossing North America, they spread across South America, where three migration routes have been identified. It is likely that certain groups also followed the Pacific and Atlantic coasts. The sea level at that time was 130 meters lower than it is today, exposing up to 200 kilometers of land on the Atlantic coast (Peltier and Fairbanks 2006). Thus, the sites of these early populations are likely to remain submerged for quite some time. The Andean route has long been considered the most ancient, but the oldest known Paleoindian sites in the Andes only date back to 12,000 years BP (Lavallée 2005); there are dates actually older than that in the lowlands (Vilhena Vialou 2005). The Amazonian route therefore seems the most ancient. It is logical to think that the groups that crossed the Central American lowlands would have preferred to continue along in a comparable environment rather than face the highland setting. Furthermore, during this period of frequent wildfires, the forest was less dominant and was replaced in many places by large patches of savanna (Bush et al. 2008; Hammond et al. 2006; Hammond and ter Steege 1998; Krook 1979; Van der Hammen and Absy 1994). Some rockshelters in Nordeste and Mato Grosso in Brazil were occupied beginning as early as 25,000 years BP (Vilhena Vialou 2005). The Amazonian rainforest as we know it today is in part the results of human activity that has continued since the Late Pleistocene. Hunter-gatherers have inhabited Guiana since at least 13,000 years BP.

The Amazon was occupied in numerous places during the Late Pleistocene and the Early Holocene. In Guiana, vestiges of Paleoindian hunter-gatherer groups have principally been found in open environments, particularly the savanna of Sipaliwini in the south of Suriname (Boomert 1980a). Multiple data indicate that, approximately 10,000 years ago, a climatic phase developed that was drier and colder than today's conditions, characterized by largely uninterrupted savannas between the coastal region of eastern Venezuela and southern Guyana, from Suriname and French Guiana (Hammond 2005; Van der Hammen and Absy 1994). Similar observations of a drier climate with open savanna during the mid-Holocene followed by an expansion of the forest in the last 3,000 years have been made in other parts of Amazonia and coastal Brazil (Pessenda et al. 1998, 2004). In Guiana, scattered, rounded hills emerge from these large open extents of grassland, covered with tall shrubs. Paleoindian populations began to exploit these areas beginning around 8000 BC (Boomert 1980a). These hunters may have burned the savannas and their edges to increase their area (Van der Hammen and Hooghiemstra 1996). These populations also may have directly exploited resources in the dense forest, but no evidence of Paleoindian sites has yet been detected in the tropical forest.

The most ancient sites in Guiana date to 13,000 years ago (Cruxent 1972). Five sites are located in Tupuken, in the Venezuelan savannas to the south of the Orinoco Delta. Large, relatively coarse tools have been found associated with wood charcoal and megafaunal remains. In Suriname, more than 20 Paleoindian sites were discovered more than 40 years ago by the forester Frans Bubberman, who traveled across the Sipaliwini Savanna several times (Versteeg and Bubberman 1992). There is no evidence of residences, but lithic production workshops are noted (Figure 26). These can be separated into two groups according to their function: occasional or specialized use. The occasional workshops, which are more numerous, indicate isolated episodes of stone-knapping, probably carried out by a hunter upon accidentally discovering a stone block that was appropriate for the manufacture of a tool. They consist of a few flakes of debitage spread over a small area, and occasionally a projectile point that was broken during its production.

The two specialized workshops cover a much larger extent, occupying approximately the area of a rugby field. The artifacts recovered include projectile points, failed tools, and those broken in the course of manufacture, debitage flakes, cores, and percussors. The size of these workshops, the variety of raw material used, and the quantity of remains present indicate that these sites were used regularly. While studying collections from

Figure 26. Paleoindian projectile point in a workshop of the Sipaliwini savanna, Suriname (photo by Frans Bubberman).

the Stichting Surinaams Museum, I classified the projectile arrow points into four principal types (Figure 27). The first is the lanceolate type, oval or triangular with straight or concave edges. The talon is straight or convex. The second is the aileron type, triangular in shape with convex or straight edges and a concave proximal edge. The third is the stemmed type, triangular in shape with convex edges and a convex stem proximal edge. The fourth is the stemmed-fin type, triangular with straight or concave edges, and a very marked concave proximal edge with a stem. The other chipped-stone tools are scrapers and oval knives.

Hunter-gatherers did not establish temporary settlements in these stone working areas (Frans Bubberman, personal communication, 2003). The caves and rockshelters found at the foot of the hills that dot the region provide more comfortable options. The walls of some of these caverns contain painted rock art, but it is not certain that Paleoindian groups created these images because these locations continued to be inhabited for several centuries afterward (Figure 28). The presence of much more recent populations is attested to by the ceramic sherds that litter the ground.

Figure 27. Paleoindian projectile points from the Barima River, Guyana (Museum of Archaeology and Anthropology at Cambridge University; photo by Stéphen Rostain), and from the Sipaliwini savanna, Suriname (Stichting Surinaams Museum; photo by Stéphen Rostain).

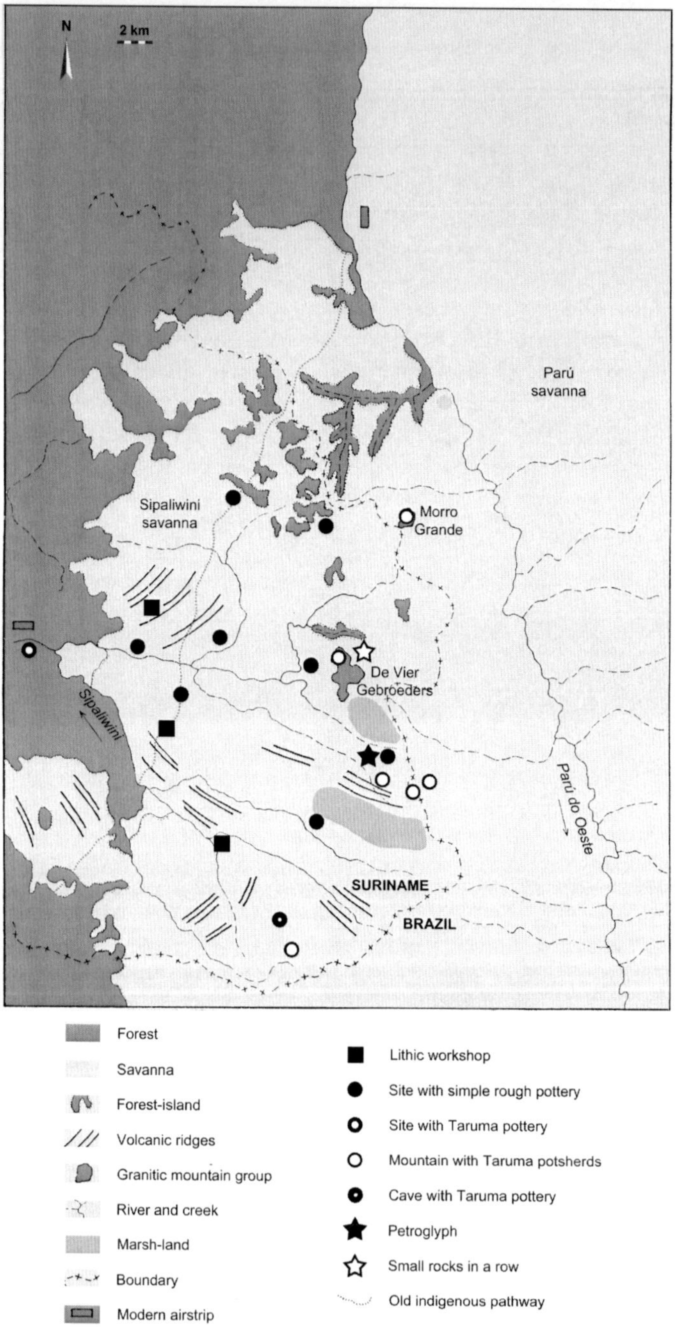

Figure 28. Archaeological map of Sipaliwini savanna, Suriname (modified from Frans Bubberman, personal communication).

The points allowed the hunting of relatively large animals on the open savannas. Two phases are distinguished in the Paleoindian settlement (Boomert 1980a): a more ancient phase of hunters of Late Pleistocene megafauna (such as mammoths, mastodons, and megatheriums), and a more recent phase of hunters of modern fauna (deer, tapirs, and other animals of this size). Thus, the first inhabitants of the Amazon hunted animals such as *Megatherium, Doedicurus*, and *Glyptodon* (giant armadillos), and perhaps *Thylacosmilus, the famous saber-toothed cat that inhabited the Amazon* (*Natural History of South America* 2010; Ranzi 2008; Veblen et al. 2007; Webb 1997).

Movement across the savannas, and hunting in particular, would not have been easy due to the covering of tall grass, often measuring more than 2 meters high, which impeded the view of the hunters' game. A simple solution that was probably adopted by hunter-gatherers would be to clear the savanna through burning. Furthermore, in an open and disturbed landscape, plant resources that could feed humans, or the animals they hunted, would occur in a higher proportion (Doyle McKey, personal communication, 2010). This repeated burning by humans would have affected the leading edge of the forest and the extension of the savannas. Some indigenous myths even attribute the emergence of the savannas to anthropogenic actions. The practice of burning has continued among indigenous societies. The Urubus-Kaapor of the Maranhão State of Brazil (Maranhonian Guiana) believe that the savannas were created by a gigantic forest fire set by the ancestor "Lune" (Ribeiro 2002). Paleoindian populations also used forest resources, in particular collecting fruits, but also for hunting, as evidenced by the discovery of isolated stone tools that were produced in the forest. Chipped stone points have also been found on the Upper Negro, close to Manaus and to the south of the Amazon on the Tapajos and the Curua.

The rhythm of occupations by Paleoindians increased rapidly. On the Orinoco, Atures-1 sites date to between 9,200 and 7,000 years BP (Barse 1990, 1995). In Colombia, the site of Peña Roja, on the banks of the Caquetá, was occupied between 9,250 and 8,090 years BP, a period when the tropical forest dominated the region. Palms represent 68 percent of the macrobotanical remains recovered (Cavelier et al. 1995); many of these species are still widely consumed by modern nomadic populations in the Amazon, such as the Nukak or the Makú (Politis 2007). For instance, Early Holocene domestication of the peach palm (*Bactris gasipaes*) has been recently demonstrated (Clément et al. 2010). On the Middle Amazon, the preceramic site of Dona Stella has produced dates from between 9,460 and 4,500 years BP (Costa 2009). In any case, around 9,000 years BP, the evidence of human presence in the Amazon multiplies.

The region of Monte Alegre on the north bank of the Lower Amazon has produced important data for the most ancient inhabitants of the Amazon. Fifty-six radiocarbon dates, spanning the period from 11,700 to 9,880 years BP, have been obtained from the Pedra Pintada rockshelter, located in the cliffs almost 10 kilometers from the Amazon River (Roosevelt et al. 1996). For several centuries, this site was visited by groups who consumed forest fruits and a wide range of terrestrial and aquatic faunal resources. They fabricated stone tools (projectile points, scrapers, and knives) and painted circles, animals, and human silhouettes on the walls of the rockshelter (Figure 29).

FISHER-GATHERERS

Around 8,000 years BP, the first semisedentary occupations of Guiana began with the settlement of groups with diets based on the exploitation of aquatic resources: fish and shellfish. While hunter-gatherer groups persisted in the interior savannas of Guiana, other populations began to settle the Lower Amazon and part of the coast. They built shell middens known locally as *sambaquís*. It has been suggested that the transition from the exploitation of the interior savannas to a coastal settlement to use marine resources could have been linked to the disappearance of megafauna game (Figuti 1992). Moreover, the modern sea level was established around 7,700 to 6,900 years BP (Arroyo-Kalin 2008; Meggers 2007).

Figure 29. Serra Paytuna rock paintings, near Pedra Pintada site, on the left bank of the Lower Amazon (photo by Serge Guiraud).

On the northern bank of the Lower Amazon, the fisher-gatherers intensively exploited fish, but they also began to cultivate plants. After having likely used gourds and waterproofed baskets as containers, they would have needed to cook some of their foods, particularly plants that could not be consumed raw. The most ancient ceramics appeared around 7,080 years BP in the Taperinha *sambaquí* close to Santarém, and even earlier in the nearby Pedra Pintada cave around 7,600 years BP (Roosevelt 1995; Roosevelt et al. 1991). At this latter site, the ceramic sequence may have persisted for 2,500 years. If this is verified, it would make the Amazon the most ancient center for the invention of pottery in the Americas, 1,000 to 1,500 years before the sites of Valdivia on the Ecuadorian coast (Meggers et al. 1965), Monsú (Reichel-Dolmatoff 1985), and San Jacinto in northern Colombia (Oyuela-Caycedo and Bonzani 2005).

The Taperinha *sambaquí* is 6 meters high and extends over several hectares. Twelve radiocarbon dates carried out on a variety of material collected in the deepest levels range from 8,025 to 7,780 and 6,665 to 6,415 years BP, indicating an occupation of more than 1,300 years (Roosevelt et al. 1991). Ceramic artifacts consist of fragments of brownish-red bowls with sandy temper, and in the upper levels, organic temper. Decorated sherds are rare (3 percent of the sample) and limited to incised rims. Lithic material consists only of hammerstones, chipped stone tools, and unworked groundstone tools. Also recovered were a bone fish-hook, mollusk and turtle shell scrapers, and a pendant made from marine mammal bone. Food remains indicate an intense exploitation of fishing and collection of fluvial resources: primarily freshwater shellfish, but also turtles and fish. Evidence of plant collection is more tenuous.

Few regions along the coast of Guiana offer significant shellfish resources. Those that do are very circumscribed by the transcoastal muddy flow of the Amazon, which extends from its mouth toward the northwest to the central coast of Guyana. These mud banks impeded the proliferation of shellfish on the coasts of Amapá, of French Guiana, of Suriname, and of eastern Guyana. In contrast, large shellfish colonies exist along the coast of Pará to the east of the Amazon and on the western coast of Guyana to the west of the Essequibo River. Fisher-collectors who colonized these two regions gathered and consumed large quantities of shellfish. Apparently, these resources supplied a sufficient quantity of protein for permanent or semipermanent settlements. Furthermore, the coastal regions are rich in other food resources, such as plants and fruits; several ecological zones could be exploited simultaneously. Crushed shell served as ceramic temper, and tools were fashioned from stone or shell.

Groups from Pará in Brazil are referred to as the "Mina" culture, while those from Guyana are the "Alaka" culture. The remnants of their villages represent one of the most impressive archaeological remains in

Guiana. Indeed, the disposal of shell from the marine animals consumed gradually formed gigantic middens that, in places, can reach several meters in height. The settlements were located at the summit of these mounds that grew progressively from one year to the next. More than 40 of these shell midden sites have been recorded in the estuary of the Amazon, the most ancient of which date to between 5,700 and 3,300 years BP (Simões 1981). The settlements of the Alaka culture in Guyana date between 6,800 and 3,550 years BP (Meggers 2007; Williams 1992).

The presence of polished stone axes in a Mina culture shell midden, dated from 3,900 to 3,500 years BP, leads one to think that they attempted agriculture. However, the Formative Mina and Alaka cultures do not provide tangible evidence of the domestication of plants. If hunter-gatherers began to manage plants very early (Politis 2001), Mina and Alaka populations may have begun to cultivate plants, but their subsistence economy continued to be based on marine resources. Nevertheless, data are available in other areas of the Amazon that suggest that slash-and-burn agriculture began to be practiced between 6,000 and 5,000 years BP (Iriarte 2007; Oliver 2008; Piperno and Pearsall 1998). Maize pollen collected in Ayauchi Lake in the Ecuadorian Amazon has been dated to 5,300 years BP (Bush et al. 1989).

However, not all of the Archaic sites of Guiana are located on the edge of the sea or a river. Some discoveries have recently been made at the interface of the coastal plain and the interior forest of French Guiana. Three Archaic and Formative sites date between the seventh and the fifth millennia BP. The sites of Carrière des Ananas and Plateau des Mines are located 12 kilometers south of Saint-Laurent du Maroni and approximately 5 kilometers from the bank of the Maroni River. These sites have provided seven radiocarbon dates spanning the period of 7,300 to 4,600 years BP (Delpech 2005; Mestre and Delpech 2008). The Eva-2 site is located on a small hill of white sand on the high coastal Pleistocene plain at the junction of the forest and the coast, close to the Malmanoury Savanna, east of Sinnamary. The first occupation of the site is dated between 6,200 and 4,000 years BP (Van den Bel 2006).

These three sites are made up of a series of small mounds of quartz pebbles, circular or elongated in shape. They contain shallow basins or holes. At Eva-2, 210 mounds have been recorded, generally distributed in regular rectilinear alignments or in half-circles. Lithic debitage and stone tools are found among these mounds and their surroundings. These lithic objects are made from hyaline quartz, whereas the gravels recovered from the mounds are poor-quality saccharoid quartz. Because fire staining indicates that these pebbles were heated, these structures have been interpreted as stone ovens (Mestre and Delpech 2008; Van den Bel 2006) and compared with the ovens used in the Andes or to cooking pits in North America used during this same period (Thoms 2003, 2008).

These structures are also very similar to those discovered in the much more recent site of Tanki Flip, on the island of Aruba (Versteeg and Rostain 1997). At this site, three types of heating pits have been recorded: (1) deep pits full of ash; (2) similar pits, but filled instead with rocks (up to 60 specimens in a single structure); and (3) large, burned pits that apparently served as ceramic kilns. In the second type of pit, blocks of limestone, granite, quartz, diabase, and diorite are found. Diorite tends to be present as broken tools. These different structures were arranged in recurring spatial layouts within houses and connected to caches and burials. This distribution suggests that they were not used simply for heating or cooking, but that they also had a ceremonial role, probably linked to funerary rites.

In contrast to the first two sites, ceramics were discovered in the lower level of Eva-2. These sherds were very eroded, with a quartz sand temper, which is similar to the ceramics at other contemporaneous Formative sites in Guiana and on the Lower Amazon.

Further inland, in the forest, a Formative site was recorded in southern Suriname. Between 2005 and 2007, a team from the Stichting Surinaams Museum studied the 303 petroglyphs of Werehpai Cave, not far from the Tirió village of Kwamalasamutu. The excavations produced ceramics associated with charcoal dated between 4,930 and 4,200 years BP (Versteeg 2007). Once again, these ceramics are coarse and poorly conserved.

It appears that the interfluvial zones of the forest were frequented by groups that were more nomadic than the coastal and fluvial fisher-gatherers. Their diet was based on terrestrial mammals and plant collection. Their use of ceramics also began later (da Silveira 1994).

It is clear that between 7,000 and 5,000 years BP, ceramics appeared in several places across Guiana, corresponding with the beginning of sedentism and the first attempts at cultivation. For example, manioc was probably domesticated beginning around 8,000 years BP in the western Amazon, in the states of Acre and Rondônia in Brazil, and to the east of Bolivia, before diffusing to the remainder of the lowlands (Iriarte 2007; Léotard et al. 2009; Olsen and Schaal 1999; Piperno and Dillehay 2008; Piperno et al. 2000). The gradual spread of these characteristics took place over a very different landscape from that of today; this period was more arid, and marked by the disappearance of large tracts of forest from fires, leaving behind circumscribed areas of forest refugia (Van der Hammen and Hooghiemstra 1996). Following the Pleistocene to Holocene climatic transition, humans progressively went from exploiting large, open areas to focusing on more forested and humid environments.

Ceramists and Cultivators

The use of ceramics and agriculture developed gradually in Guiana. By 4,000 years BP, humans first began to construct true villages, as well as ceramics and the tools necessary to cultivate fields, which were created by clearing the forest with axes and fire. These cultivators dominated their territory, crossing it either on foot or by canoes on its rivers. It is likely that during this period they lived in small villages, similar to modern indigenous settlements. As the environment degraded, the group would move their village to a more profitable location. Thus, the landscape became increasingly exploited and transformed through time.

It was also during this period that true agriculture emerged. Like Mesoamerica and the Andes, the Amazon was a center for plant domestication. Its principal cultigen, manioc, is unusual because it is a poisonous tuber. A complex series of steps are necessary to render bitter manioc edible, but thanks to a sophisticated technology, indigenous peoples were able to transform an extremely toxic plant (as well as its nontoxic variety) into multiple food products, including flour, flat cakes, sauces, and beer (McKey and Beckerman 1993). Although domesticated for several millennia, manioc began to be intensively cultivated only during this period, with the use of sweet manioc perhaps preceding that of the bitter variety (Arroyo-Kalin 2008, 2010; Clément et al. 2010).

Agriculture developed progressively with the population's diet based primarily on domesticated plants. At least five principal agricultural techniques were practiced in the Amazon during the pre-Columbian period (see Chapter 1). The first was agroforestry, which involves favoring the development of selected plant species in cleared areas. The second was slash-and-burn agriculture, in which several trees are felled to open a space for planting. The third technique is conducted on *terra firme* and is carried out on the highly fertile anthropogenic black soils known as *terra pretas* (Arroyo-Kalin 2008; Glaser and Woods 2004). The final two techniques were practiced in seasonally flooded areas. *Várzea* agriculture was carried out during low-water periods using the rich alluvial sediments that are deposited annually along the Amazon and the Orinoco. Finally, the raised-field technique was amply applied to the seasonally flooded savannas of the Amazon and in Andean valleys (Denevan 2001; Erickson 2008; Rostain 2008a, 2010a). The last three intensive techniques developed and diffused throughout the Amazon at the beginning of the Common Era.

An increase in sedentary farming villages and the emergence of more structured societies occurred during the Formative period (4,500–2,000 years BP). Beginning around 4,000 to 3,500 years BP, sites with elaborate ceramics began to appear throughout the Amazon. Some groups chose the rich volcanic soils of the Andean piedmont, where settlement began at 4,000 years BP (Valdez 2008b). Similarly, artificial earthen mounds were constructed in the Upano Valley, Ecuador, beginning in the first millennium BP (Rostain 1999a, 2012).

Other sites were located in the floodplains that border the large rivers. In Venezuela, the Agüerito-Cedenoid tradition of the confluence of the Orinoco and the Apure is dated from 5,300 to 4,000 years BP (Lathrap and Oliver 1987; Zucchi and Tarble 1982; Zucchi et al. 1984). In Peru, the Tutishcainyo Hatched-and-Incised tradition of the banks of the Middle Ucayali is estimated to date to 4,000 to 3,800 years BP (Lathrap 1970; Oliver 2001).

In Guiana, the site of La Gruta-Ronquin, on the Middle Orinoco (2600–1000 BC), has produced ceramics from the Saladoid tradition. These are characterized by a very careful decoration of geometric motifs painted in white on a red background, and zoomorphic models with broad incisions applied to the rims (Cruxent and Rouse 1958–59; Vargas Arena 1979, 1981).

The Saladoid tradition expanded remarkably beyond its original hearth in the Lower and Middle Orinoco beginning in the final millennium BC, spreading to the Lesser Antilles, where it became most common. It also expanded to the east, as far as western Suriname, resulting in settlements along small tributaries, such as that of Kaurikreek. This site extended over a hectare on a sandy bank on the edge of the creek, facing a swampy plain. A piece of wood charcoal recovered in excavation dates from 2200 to 1750 BC (Versteeg 1978). The ceramics are characterized by an unusual temper, made of laterite, chalcedony, kaolin, and quartz (Versteeg 2003) and a decoration made by applying fine cords in a net pattern to the body of the vessel, recalling basketry motifs (Rostain 2009a). Highly stylized and incised zoomorphic models complete this decorative style. Eastern Guyana also has yielded very ancient Saladoid sites with ceramics similar to those of Kaurikreek (Whitehead et al. 2010).

The site of Wonotobo was established on a small hill 200 meters from the Middle Corentijn in Suriname, not far from rapids. It has been dated between AD 70 and 200 (Boomert 1983). Apart from the wide-stemmed handles, the decoration is different from that of Kaurikreek, and white sand is used as temper. The dominant decoration consists of rounded, parallel motifs in red on a white background (or the inverse). Also present are fine parallel or lattice incisions surrounded by wider incisions.

These Surinamese sites of the Saladoid tradition were linked to rivers away from the coast. No Saladoid settlements have been found east of the Corentijn River, which thus represents the eastern limit of this tradition in Guiana. Although villages multiplied during this period, there remained many unoccupied areas of the Amazon. Thus, these populations still maintained large territories.

Farmers

At the beginning of the Common Era, settlement density increased in the Amazon's two large fluvial basins. The fertile banks of the Amazon and the Orinoco attracted farming peoples. This demographic expansion resulted in the displacement of communities that spread along the coast of Guiana. This Neolithic revolution began around AD 300 and was fully under way beginning around AD 600. The riverbanks and the coast were settled by populations that managed and transformed their environment.

The sociopolitical structure of these communities evolved with the emergence of hierarchy, the specialization of certain tasks, the organization of an exchange system, and the development of shared ceremonial practices. On the coast, cultural spaces became more defined, generally delimited by major rivers. The population increase resulted in a more sensitive distribution of territories and more assertive cultural identities. These people not only occupied short-term villages, but they also formed large aggregations of several hundred individuals in long-term settlements.

By AD 500 to 600, sites began to increase in the fluvial and maritime floodplains. Some researchers consider this an indication of population growth. What is certain is that during this period, the two large fluvial basins and the Guianese coast were densely settled. Important cultural traditions developed initially along the Amazon and the Orinoco before spreading to other regions (e.g., Incised-and-Punctate, Polychrome, Barrancoid, and Arauquinoid). The growth of large populations with complex social structure gradually forced small-scale cultivators of *várzea* along major rivers and the coast to move into the interfluvial regions that they continue to inhabit today. Thus the *várzea* was highly settled, and competition for arable land would have provoked conflict. Enormous villages spread over several kilometers along the riverbanks.

Around AD 600, four successive cultures belonging to the Arauquinoid tradition (Hertenrits, Kwatta, Barbakoeba, and Thémire) appeared on the western coast of Guiana, originating from the Middle Orinoco. At the same time, eastern French Guiana and Amapá were occupied by groups that originated from the Lower and Middle Amazon. Later, around AD 1000, the space was colonized by the Late Aristé culture, linked to the Polychrome tradition of the Lower Amazon. At this time, the Guianese coast was under the influence of two major cultural hearths: the Arauquinoid tradition of the Middle Orinoco to the west, and the Polychrome tradition of the Lower Amazon to the east (Rostain 1994b, 1994c).

The Cultural Hearth of the Lower Amazon

The Middle and the Lower Amazon represent the most remarkable center for the development of Amazonian cultural traditions, from which various groups were born. The most prolific among theme was the Polychrome tradition, which was associated with at least 15 different cultural entities. They ranged from the mouth of the Amazon to the foot of the Andes, with extensions in Peru and Colombia.

The Polychrome Tradition

The Polychrome tradition developed in the Lower Amazon just after the beginning of the Common Era. This tradition diffused widely. It is not only found in the piedmont of the Andes in Bolivia, Peru, Ecuador, and Colombia; it also occurred on the Atlantic coast from the mouth of the Amazon, north to Île de Cayenne. The farther one goes upstream, beginning at the mouth of the Amazon and moving toward the Andes, the more recent are the sites of the different Polychrome cultures, indicating that the Polychrome tradition originated in downstream reaches of the river (Neves 2006). This Polychrome tradition flourished in particular on the left bank of the Lower Amazon, where a variety of related cultures developed simultaneously. The profusion of related Polychrome cultures in the state of Amapá and the islands at the mouth of the Amazon is astonishing and unparalleled elsewhere in the Amazon. It appears that large complex societies flourished on the ocean shore and along the river around AD 1000.

One of the most remarkable ceramic vessel types was the richly decorated anthropomorphic funerary urn. These urns were used to contain the bones or ashes of the deceased, and were then buried, or more frequently deposited in a rocky cavity. Grouped in an open-air cemetery, a cave, or artificial pits, the urns represent an ensemble of transformed beings that recreates a closed microcosm of ritual meetings comparable with those of living people. All indications are that despite a change in their state, the community life and cultural identity of the dead mirrored those of the living. Such practices seemingly reflect a desire to continually perpetuate the image of the society. Even in death, existence was above all social, based on the recognition of others and not on the dynamic of the individual. In the 1950s, the Carajá of the banks of the Araguaya River in Brazil still practiced secondary burials, exhuming human remains and depositing them in urns called "little bone houses" (Lelong 1953), symbolizing the reconstructed image of the society in death.

The culture most representative of the Polychrome tradition is the Marajoara, which developed on Marajó Island from AD 450 to 1350 (Meggers and Evans 1957; Nordenskiöld 1930; Roosevelt 1991; Schaan 2004). Following a local development, the Marajoara culture emerged in the seasonally flooded savannas to the east of the island, where it occupied artificial earthworks. Marajoara ceramics are characterized by elaborate curvilinear or rectilinear geometric motifs painted in red, black, brown, orange, and white and executed with complex incisions and excisions, as well as original and elaborate vessel forms.

Four cultures developed side by side in the south of Amapá. Three are linked to the Polychrome tradition. The final, Mazagão, is related to the Incised-and-Punctate tradition. The north of the state of Amapá and eastern French Guiana were dominated for more than a millennium by the Aristé culture, the best known in the region.

The Maracá culture. Established in the center of the Mazagão territory, the Maracá culture seems to have appeared slightly before European contact. These sites were distributed along the small Maracá tributary that flows into the Amazon. Although discovered at the end of the nineteenth century, this culture was not subject to archaeological excavation until a century later; thus, it remained poorly understood. The domestic settlements, distributed along a small river, did not extend far, between 60 and 200 meters long, and are dated by a single radiocarbon date of 360 ± 40 years BP (AD 1445–1645, calibrated) (Guapindaia 2001).

Cemeteries were located in small caves or rockshelters in the hills close to the settlements. Funerary urns were deposited in groups (Figure 30). The individual vessels depict either men or women seated on a bench, or four-legged animals. In some cases they are simple, undecorated tubular containers. The polychrome paintings depict elaborate motifs dominated by straight lines. Secondary urn burials do not show evidence of cremation, but rather of bones recovered after the decomposition of the body.

European glass beads dated to the seventeenth century incorporated in the décor of one urn indicate that this culture persisted during the colonial period, and had some degree of contact (even if indirect) with westerners traveling along the Amazon. The location of these groups upstream along an inconspicuous tributary probably protected them from the first colonists. However, they likely disappeared during the course of the eighteenth century.

The Caviana culture. This culture occurred on the eponymous island and the coast of Amapá, and has not yet been precisely defined. Nevertheless, it without a doubt represented a culture entity in its own right, distinct from its neighbors. Excavations carried out in 2007 in a Macapá funerary pit will likely shed light on the identity of this culture (João Darcy de Moura Saldanha, personal communication, 2008).

The study of ceramic pieces in museums and private collections show a specific style. Nonetheless, only funerary urns have been studied to date (Rostain 1996). They are tubular, more or less globular, representing a man or a woman with clear sexual attributes, frequently seated on a bench. The head generally serves as a cover, and the arms and legs are generally well marked in relief. Some pieces in the round are extremely realistic. Very delicate spiral motifs in black, brown, or red paint cover the entire surface of the vessel. The style of the painting and the ceramic form borrows as much from the Late Aristé and Maracá cultures as it

Figure 30. Cave with funerary anthropomorphic urns, Maracá culture, Amapá State, Brazil (photo by Serge Guiraud).

does from the Marajoara, but this creates an original aspect. By analogy, this culture could be dated to between AD 1000 and 1500.

The Aruã culture. The territory of the Aruã culture is limited to the Mexiana and Caviana Islands in the Amazon estuary, with isolated intrusions to the north of Marajó Island and on the southern coast of Amapá, where it is found within the Aristé sites. Due to the crude aspect of its ceramics, it was first thought to be the most ancient ceramic culture of Amapá (Meggers and Evans 1957). However, absolute dating carried out on the Aristé culture demonstrates that the Aruã were later (Rostain 1994a). Therefore, it seems that the Aruã culture could be likened to historic groups of the same name that occupied this area. Aruã funerary urns are tubular with a human face sculpted on the neck and an opening that is oriented toward the side. They may also instead depict animals.

The Aristé culture. Beginning in AD 600, eastern Guiana was occupied by communities that developed originally in the Lower and Middle Amazon. The first groups belonged to the Early Aristé culture, apparently tied to the Incised-and-Punctate tradition of the Lower Amazon (Meggers and Evans 1957; Rostain 1994a). They spread to Amapá, giving rise to two different cultural developments, the Early Aristé and Mazagão (Meggers and Evans 1957). The latter group settled to the south of the Araguari River, while the Aristé culture colonized the north of Amapá. Although the two groups experienced different developments, they maintained exchange relations. Based on studies of ceramics, correlated with stratigraphic and chronological data, I recently revised the previously established Aristé sequence (Rostain 1994a) to distinguish three successive phases: Early Aristé (AD 600–1100), Late Aristé (AD 1100–1600), and Terminal Aristé (AD 1600–1750).

The pottery was initially tempered with quartz sand or crushed shell, which was later replaced by crushed ceramic material. Domestic vessels were decorated principally with broad incisions, punctuations, and notches that formed simple motifs. In contrast, funerary ceramics were very elaborate and evolved significantly through time. Originally simple globular jars with vertical necks, the urns later took on more sophisticated forms, similar to the human body. Some models suggest faces, arms, and legs. Very refined motifs, including often-mirrored double spirals, were painted in red, brown, or black on a white background (Figure 31).

At the beginning of the sequence, during the Early Aristé, an exogenous style, named Caripo Kwep, appeared in Oyapock Bay (Rostain 1994a). The ceramic, tempered with *cariapé* (also known as *kwep,* which is burned and crushed bark), presented a completely new decoration of convoluted motifs described by broad incisions and punctuations. They resemble very much

Figure 31. Megalithic site of Rego Grande, Aristé culture, Amapá State, Brazil (photo by Stéphen Rostain).

the Itacoatiara (Hilbert 1968) and Açutuba style ceramics of the Middle Amazon, dated from 300 BC to AD 400 (Lima et al. 2006). If the appearance of Caripo Kwep resulted from the arrival of new immigrants, they must have incorporated themselves peacefully into the local populations, as no other major changes have been observed to accompany this stylistic intrusion. Fusion and assimilation are classic mechanisms in Amazonian intertribal relations (Whitehead 1993, 1994, 1996).

The Aristé occupied the hills that emerge in these swampy plains and the high banks of rivers, between the Approuague in French Guiana and the Araguari of Amapá, an area 400 kilometers long and 10 to 100 kilometers wide. They were occasionally present on Île de Cayenne. Their villages were modest in size, 2,000 to 40,000 square meters, and they practiced a lifestyle closer to cultivators at the beginning of the Common Era than that of specialized farmers in western Guiana and the banks of the Amazon. These settlements of several families contrast with large villages that housed larger communities, comparable with those observed on the Lower Oyapock by Europeans in the sixteenth and seventeenth centuries. In the hills of Oyapock Bay, rockshelters with an average area of 28 square meters were temporarily occupied outside the villages. These may have served as places for occasional retreats for individuals or families, such as for the seclusions imposed by puberty, the training of shamans, the initiation of chiefs, births, the *couvade*, mourning, or various manufacturing activities (Rostain 1994a). Cemeteries consisted of rows of urns, directly on the ground or buried, in caves or artificial pits with lateral chambers closed by granite slabs. Secondary burials characterized the Early Aristé culture, which were later replaced by cremation and grouping of ashes in urns during the Late and Terminal Aristé. Ceramic offerings were numerous, and during the colonial period, many European artifacts and glass beads accompanied the burials. The presence of melted beads amalgamated with the bone indicates the deceased were burned with their possessions or with offerings.

Remarkable sites occur in northern Amapá, built from enormous upright slabs that are arranged in circles or in lines on top of small elevations. Originally attributed to the Aruã culture (Meggers and Evans 1957), they were actually constructed by the Aristé (Boomert 1981; Rostain 1994a). Excavations and observations carried out over three years at the megalithic site of Rego Grande during the winter solstice (Figure 31) have revealed correlations, particularly between the alignments of upright stones at the hill's summit and the sun at certain hours of the day (Petry Cabral and Darcy da Moura Saldanha 2009). It appears that beyond their ceremonial function, they also served as cemeteries. These arrangements of upright stones are similar to the megaliths and giant statues used for fertility, marriage, and birth rites in the central Amazon during the seventeenth century (Nordenskiöld 1930).

The preponderance of ceremonial sites (rockshelters, cemeteries, megalithic groups) over domestic sites is a salient characteristic of the Aristé culture. It is possible to propose an organizational model based on archaeological survey and the specific distribution of the sites. This model is quite clear in the hills of Oyapock Bay. The villages were arranged on the southern slope, not far from a creek. The rockshelters that were temporarily occupied by one or two individuals were located further uphill. At the summit of the hills, far from the habitation sites, cemeteries of urns in caves are found. It is likely that, similar to many modern indigenous groups, these pre-Columbian peoples did not cohabit with the community of the dead. Instead, the dead were confined to a circumscribed space. The cemeteries were thus isolated to maintain the stability of the settlement of the living, allowing a veneration of ancestors that was linked to a precise place. The division of residential and ritual sites reflects the importance of ceremonial life in the Aristé world.

The number of habitation sites, the high density of the necropolis and occupied rockshelters, and the diversity of the local ceramic innovations suggest that the Lower Oyapock represented a particularly important focus for Aristé development. The presence of coastal hills and their strategic advantage played a stabilizing role for the population. At the beginning of the colonial period, the chiefs of the pantribal confederations of Amapá resided in this location, which was apparently a crucial political center.

The Incised-and-Punctate Tradition

A pocket existed in the Lower Amazon where the Polychrome tradition was replaced by very different cultures, linked to the Incised-and-Punctate tradition. Indeed, they represented the Amazon's counterpart to the Arauquinoid tradition of the Orinoco and of Guiana, although it is not yet known if they were independent developments or if one preceded the other. Although the ceramics show differences in form and decoration, their styles are generally similar. Incisions, punctuations, and reliefs are comparable.

The Santarém culture is dated from AD 1000 to 1600, up to European contact. The villages are between 2 and 3 kilometers in diameter, and would have contained 10,000 to 20,000 inhabitants. The archival sources from the sixteenth and seventeenth centuries describe the presence of a powerful indigenous community at the mouth of the Tapajós River, the principal settlement located at what is now the modern city of Santarém. Santarém ceramics are of an extremely baroque style with elaborate forms, from vases to caryatids and realistic anthropomorphic figurines in various postures, often acrobatic. The texts of the first explorers that encountered these societies indicate that women played important political and religious roles.

Other essential artifacts associated with this society, green stones (Amazon stones, or *muiraquitãs*) are found throughout the Amazon and the Lesser Antilles. Most of these zoomorphic pendants depict frogs, and less frequently, birds, fish, turtles, nonidentified animals, or geometric figures. Indigenous populations have always manifested a great interest in these green stones. They would have been exchanged with the Amazons, "the women without men," for colorful parrot and macaw feathers, typically masculine attributes in Amazonian societies. Green stones were feminine adornments and the symbols of myths that were associated with feminine connotations, as they contain references to water, lakes, softness, the color green, darkness (night), the anaconda, and the moon, as well as animals linked with women, such as frogs, lizards, and fish (Wassén 1934). Indigenous peoples believed that the *muiraquitãs* were fashioned from lake mud that hardened like stone on contact with the air. The production and exchange of green stones began around 400 BC and continued until the beginning of colonization (Boomert 1987; Rostain 2006b). Four zones of green stone manufacture are known: the Lower Amazon (Palmatary 1960), the central coast of Suriname (Boomert 1987), around Valencia Lake on the coast of Venezuela (Antczak and Antczak 2006), and the north of the Lesser Antilles. *Muiraquitãs* constituted the principal means of intra- and interethnic ceremonial exchange. Bartering took place between tribal chiefs for marriage or peace transactions, for example, with these green stones representing the currency.

Stone statues found along the Amazon are generally attributed to the Konduri culture, which is related to the Santarém culture. They often represent a human being held by an animal from behind. No iconographic interpretation of these pieces has yet been attempted, but I propose a hypothesis. The cavity found behind the statues could have served as a vehicle for inhaling or ingesting hallucinogenic plants, a common practice in the Amazon. The representation of a carnivore behind a human recalls the visions provoked by certain hallucinogens such as *takini* (*Brosimum acutifolium*, Moraceae) (Grenand et al. 2004). The latex of the tree is ingested during the shamanistic rites of several ethnic groups in French Guiana, causing painful muscular contractions and the sensation of fever. Palikur shamans and the Wayampi of Guiana say that this practice allows them to sense the presence of the tree spirits behind them, which appear in the form of a roaring jaguar or sometimes a monstrous blowing caterpillar. The shaman must remain motionless because he risks death if he turns to see the spirit (Grenand et al. 2004).

The representation of a feline (probably a jaguar) is found on several wooden snuff trays of the Maué of the Amazon (with an exceptional example where the animal is behind a woman (Torres 1987; Wassén 1965). The Konduri pieces are exceptional, although they resemble the San Augustín statues discovered in the Andes of Colombia. These statues, perhaps of the same type, were still used at the beginning of the seventeenth century.

While on an expedition on the Upper Oyapock in French Guiana in 1608, Robert Harcourt observed that "they possessed a stone statue that was worshipped like a god. They placed it in house that had been constructed in its honor and maintained with great care. The image of this god had the form of a man squatting on his heels, with his knees spread apart, and his elbows resting on them" (Harcourt [1613] 1926).

The Mazagão culture. The Mazagão culture developed to the south of the Araguari River in Amapá. It seems to have appeared in the first centuries of the Common Era. It evolved regularly until the end of the sequence, when a number of new ceramic styles suddenly appeared. It then disappeared during the colonial period, apparently shortly after 1630.

The Cultural Hearth of the Middle Orinoco

Similar to the Amazon, powerful cultural traditions developed along the Lower and Middle Orinoco. They played an essential and founding role throughout the Amazon, and also in the Caribbean Islands.

The Barrancoid

The Barrancoid tradition developed in the same region as the Saladoid tradition and it expanded similarly in all directions. It may even have reached the foothills of the Andes, on the Ucayali River in Peru (Lathrap 1970). It appeared around 900 to 700 BC and continued in several regions until AD 700 to 800 (Sanoja Obediente 1979).

The Barrancoid tradition replaced the Saladoid, perhaps even in the same locations, along the Guianese coast, into western Suriname. It was represented by the Abary and Mabaruma cultures. In the site of Wonotobo, it appears superimposed over the previous Saladoid occupation (Versteeg 2003). The ceramics are similar, although the newer forms and decorations are of better quality. The characteristic rims are thick and outcurved horizontally. They are decorated with very broad parallel incisions that are straight or often curved. Decorated handles are common. Rounded protuberances incised with spirals are a diagnostic feature. Anthropozoomorphic forms in relief frequently occur, and although it is sometimes possible to recognize an animal (such as the tapir), many more appear to be mixed entities, neither completely human nor animal. They instead represent images of the forest spirits.

In Suriname and Guyana, the first raised fields (small and quadrangular) were constructed by Mabaruma groups (of the Barrancoid tradition), who also constructed the settlement mounds of Buckleburg-1 and -2 (Versteeg 1983). The earliest date obtained from the Buckleburg-1 mound is 1,845 ± 55 years BP, while that for its abandonment (dated from the most recent occupation level at the mound's summit), occurs at 1,315 ± 55 years BP (Versteeg 1985). The ceramics are typically Barrancoid and tempered with quartz, *cariapé*, crushed shell, or mica. Excavation has revealed remains of deer, caiman, catfish, and birds that were consumed, evidence of the exploitation of several environmental zones (Versteeg 1985). Like their predecessors of the Saladoid tradition, the Barrancoid do not appear east of the Nickerie River. Nevertheless, I have observed some decorated sherds in French Guiana that resemble Barrancoid types. Thus, it is not inconceivable that one day this tradition will be discovered farther east than its current recorded distribution.

The Arauquinoid

Increase in the surface area of raised fields in the lowlands of Suriname, Venezuela, and French Guiana correspond to increases in population of the Arauquinoid tradition, originating from the Middle and Lower Orinoco. These communities displayed a strong cultural homogeneity, marked by several local variants, particularly in ceramic production. Several Arauquinoid cultural complexes coexisted and succeeded one another along the Guianese coast (Rostain 2009b; Rostain and Versteeg 2004).

The cultural hearth of the Arauquinoid tradition is found on the Middle Orinoco, at its confluence with the Apure. The sites spread along the high banks upstream on the tributary and downstream along the major river (Cruxent and Rouse 1958–59; Howard 1947; Petrullo 1939; Roosevelt 1980, 1997; Rouse and Cruxent 1963; Zucchi 1975). Paradoxically, this tradition remains only vaguely defined in Venezuela, due to the habit of creating a distinct cultural group (even a tradition) for every site excavated, which hinders overall understanding of ancient settlement patterns. Irving Rouse and José Cruxent started this approach, designating a style for every site. These styles could be combined or superimposed on one another. As a result, the Arauquinoid tradition is sometimes considered to have fused with the Cedeñoid, Osoid, and Valloid traditions of Venezuela (Zucchi 1975, 1991; Zucchi and Denevan 1979; Zucchi and Tarble 1984).[1] Comparing these different ceramic groups, it appears that they could belong to the same Arauquinoid tradition, and that their distinctions only represent differences among members of the same general cultural development. Based on the decorated ceramics, two major groups have been distinguished in the Arauquinoid tradition of Venezuela (Rostain and Versteeg 2004). The Apure–Middle Orinoco subtradition links styles from the Caño Caroni, Matraquero, Arauquín, Camoruco, and Corozal sites, whereas the Lower Orinoco subtradition links the Guarguapo and Apostadero styles. The former is the more ancient, with dates that vary between AD 400 and 600. The Arauquinoid tradition continues fairly late on the Orinoco, perhaps even up until European contact. The site of Tucuragua on the Middle Orinoco has provided a date of AD 1385 (Arvelo 1978).

The Arauquinoid tradition spread progressively from its origin in the Apure–Middle Orinoco region eastward along the Guianese coast beginning around AD 600. Between eastern Guyana and Île de Cayenne, it produced six individual cultures with clearly shared essential characteristics. From west to east, these are the Apostadero, Late Mabaruma, Hertenrits, Kwatta, Barbakoeba, and Thémire cultures. The first two are poorly understood and cannot be precisely characterized. The absence of Arauquinoid sites between the Orinoco Delta and eastern Guyana has various explanations. Little archaeological work has been carried out in this region because it is principally only the *sambaquís* sites that have attracted the attention of archaeologists. This bias in excavation has left an important blank on the map. Nevertheless, at the Museum of Archaeology and Ethnology of Cambridge, I found a cylinder stamp with typically Arauquinoid decorations and paste that was recovered from the Barima River, in western Guyana. The second reason is perhaps the role of modern construction works, particularly the on the polders, which likely have hidden a number of archaeological sites. Finally, the Arauquinoid populations themselves may have not settled these portions of the coast, because it is not clear that they would have left the Orinoco using a coastal route. Indeed, during the colonial period, inland routes were preferred for trade. The most favored route originated on the Orinoco via interior paths that followed the Essequibo River and that finally ended on the central coast of Guyana (Arvelo-Jiménez and Biord 1994). Therefore, it is possible that ancient Arauquinoid communities may have followed the same routes before turning east to Île de Cayenne.

These cultures belonged to the same Arauquinoid tradition and should be considered a geographic, chronological, and cultural continuum. Beyond their common chronological distribution, the traits that are most frequently associated with Arauquinoid cultures in Guiana are (Rostain 2003):

- The establishment of villages on the coastal plain on *cheniers* or artificial mounds.
- Agriculture on raised fields associated with drainage systems.
- Specialization by certain villages in the manufacture of tools and artifacts, trade with foreigners, intensive agriculture, or ceremonial life.

- Cemeteries of urns located within settlement sites.
- Cross-cultural exchanges of products within a trade network.
- Common ceramic styles with recurring forms and decorations.
- The appearance of clearly anthropomorphic figurines and appliqués, in contrast to the previous Barrancoid tradition where these representations were hybrids.
- Similar ceremonial artifacts such as green stone pendants depicting amphibians, double zoomorphic appliqués, horned ceramic objects, and pregnant female figurines.

The Hertenrits culture. The most ancient Arauquinoid culture on the coast of Guiana dates to approximately AD 600 (Figure 32). The Hertenrits culture occurs in eastern Guyana and western Suriname between the Berbice and Coppename Rivers, which represents an area close to 200 kilometers long and 25 kilometers wide. The coastal plain of Guiana is widest in this region, with few *cheniers* in its large marshes.

The Hertenrits culture is unique among the cultures that make up the Arauquinoid tradition of Guiana because their settlements are focused on a marshy segment of the recent coastal plain that contains practically no areas that are not submerged. To the east of the Coppename River, the morphology of the coast that was settled by other Arauquinoid cultures is distinct, with abundant *cheniers*. The Hertenrits populations were forced to construct clay mounds above the flood level to build their villages, continuing the method established by their Barrancoid predecessors in the same location. The stratigraphy of the Arauquinoid mounds of Hertenrits and Wageningen-1 shows similar construction processes. The levels were superimposed gradually, with dark occupation layers succeeding lighter levels of construction fill (Versteeg 1985). Dates from the soils are bracketed between AD 600 and 950, depending on the elevations of the mounds, which measure 2.9 and 2.25 meters high, respectively. Their surroundings were also modified with sunken roadways, canals, and raised fields.

The few raised elevated areas of land in the region were also occupied, and some reserved for ceremonial activities. The site of Prins Bernhard Polder, formed by small, artificial earthen mounds, is located in a distant swampy zone (Versteeg 1985). The exceptional quality of its ornaments, richly decorated ceramics, elaborate artifacts, and human figurines suggests it was used for very specific functions, most likely ritual (Figure 33).

Various excavations have revealed a great diversity of terrestrial and aquatic faunal remains, indicating a rich diet. Deer (*Odocoileus virginianus* and *Mazama americana*) and caiman (*Caiman crocodilus*) dominate, followed by manatee (*Trichechus manatus*), catfish (Ariidae) and crabs (*Ucides cordatus*). Also found are snakes, which may include

Figure 32. Hertenrits pottery from the Hertenrits mound, Suriname (Stichting Surinaams Museum; photo by Stéphen Rostain).

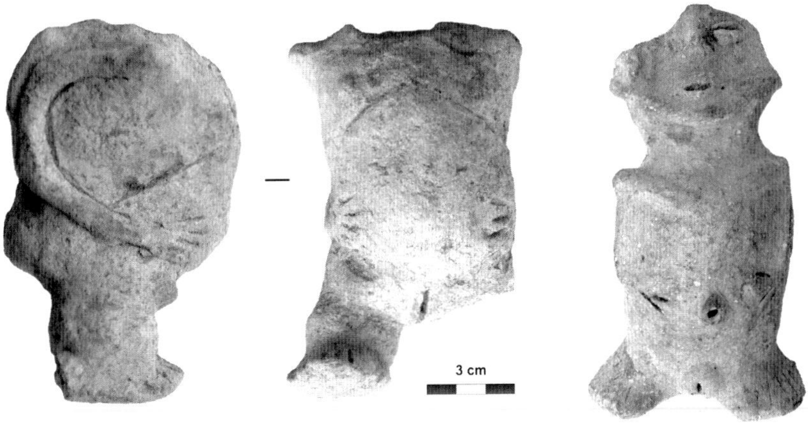

Figure 33. Arauquinoid figurines of pregnant women from the Prins Bernhard Polder site (Stichting Surinaams Museum; photo by Stéphen Rostain).

anaconda (*Eunectes murinus*), terrestrial (*Geochelone* sp.) and aquatic (*Kinosternon scorpioides*) turtles, electric eels (*Electrophorus electricus*), and different bird species: storks (*Mycteria americana*, *Jabiru mycteria*, and *Euxenura maguari*), flamingos (*Phoenicopterus ruber*), and Muscovy duck (*Cairina moschata*) (Versteeg 1985). Monkeys, peccaries, and iguanas are absent from the faunal assemblages, a finding that is puzzling because these animals are widely consumed by modern indigenous populations.

Funerary rites were performed on these mounds, where different types of burials occurred: primary burials, secondary burials in urns, and mixed primary and secondary burials. At the Hertenrits mound, 10 primary burials contain four males and six individuals of indeterminate sex, measuring 1.6 to 1.65 meters tall, in an extended position or with flexed legs, and the face turned toward the west or the southwest (Geijskes 1964). The burials found contained one or two ceramic offerings, one of which contained bird bones. Two crania were placed in urns, one of which was closed with a lid. The mixed burial contained the skeleton of a male in an extended position. His head was separate from his body and covered with a ceramic vessel (Boomert 1980b). At the Wageningen-1 mound, the remains of a robust male individual were recovered. He is estimated to have been between 40 and 70 years old, with an intentionally deformed cranium and molars with abundant caries. The dispersed remains of at least two other individuals are present as well, one of which is a man over 18 years old (Versteeg 1985). Finally, at the site of the Peruvia-2 *chenier*, an urn held preserved burned human bone from an individual estimated to be between 40 and 70 years old.

In two other urns were found the remains of a child (probably a boy), 9 to 10 years in age. He appears to have suffered physical hardships. One of these urns also contained bones from an individual who was at least 14 years old. Other burials presented the scattered remains of a child estimated between 3 and 8 years old, and of at least four other persons, of which two were men and two were women (Versteeg 1985).

The Early Hertenrits ceramics began as relatively simple, rough vessels, with incised lips and winding cord appliqués. They later became more elaborate and diverse, particularly beginning around AD 1000, during the Late Hertenrits period. Principally tempered with crushed ceramic, the vessels displayed many open forms. Relief is often present on both sides of the rims of plates, sometimes forming marked stems or twin stems. Decoration is composed of serrated lips, punctuated or undulating horizontal cords (sometimes in spirals), rims with interior punctuations, and parallel incisions in chevron patterns. Flattened vertical handles made from parallel ridges are common. Others vertical handles are flattened and incised with two notched lines that continue along the body.

Small modeled projections, often in pairs and extending beyond the rim, often seem to represent hybrid human/animal figures. The appearance of clearly anthropomorphic representations in Middle Orinoco sites is interpreted as a sign of the emergence of social stratification and the veneration of ancestor-chiefs (Roosevelt 1997). Figurines of pregnant women are frequent in Arauquinoid cultures (Figure 33). It is clearly tempting to associate this type of representation with an agricultural fertility cult in a society where the diet clearly depended on the harvest from the raised fields.

Finally, it is important to point out the presence of a series of enigmatic ceramic objects, probably for ceremonial use: tubular artifacts with two horns or discs with a central knob. Shell was used to manufacture discoidal, rectangular, or elongated beads, earrings, small labrets, etc. Tools were made from the bones of terrestrial mammals and of caimans, and from turtle shell (Versteeg 1985).

Three Arauquinoid cultures that overlap chronologically ranged along the Coppename River to Île de Cayenne, establishing villages on sandy bars and occasionally on the high banks at the mouths of rivers. Apart from their distinct ceramic styles, each of the cultures seems to have practiced complementary economic activities. Exchanges were facilitated by the linear distribution of the settlements, with goods transported by maritime or overland routes that connected the villages that were arranged on the *cheniers* parallel to the shore.

It is possible to characterize broadly the specializations of the three Arauquinoid cultures from eastern Guiana, called the Kwatta, Barbakoeba, and Thémire (Rostain 2010c).

The Kwatta culture. The Kwatta culture was located between the Coppename and Suriname Rivers, in a space approximately 100 kilometers long and 30 kilometers wide. Radiocarbon dating places the settlements between 1,140 and 895 years BP (Versteeg 2003). Unlike other Arauquinoid cultures of Guiana, these people did not build raised fields. This may be explained by the fertility of the series of sandy shell ridges that follows the central coast of Suriname and makes slash-and-burn agriculture sufficient to meet the needs of these populations. To the south of Paramaribo in Suriname, Kwatta-Tingiholo, the largest site of this culture, was established 1,350 to 1,000 years ago on a *chenier* that was formed by maritime currents 1,715 years ago. Moreover, their neighbors to the west, of the Barbakoeba culture, could have provided crops from their raised fields. The Kwatta groups depended little on intensive agriculture, instead specializing in the production of prestige objects and developing a trade network to obtain their raw materials and to distribute their finished products. In addition to technologically very elaborate ceramics, they produced a great variety of tools and, above all, sets of stones, shell, bone, and probably others from perishable materials.

Their most prestigious production was no doubt the *muiraquitãs*. These polished green stone pendants that represent fairly stylized frogs are among one of the most popular and prized items of indigenous jewelry in the Amazon. Close to 50 *muiraquitãs* were discovered at Kwatta-Tingiholo, which was one of two large Amazonian production centers for these valuable pendants, the other located on the Middle Amazon around Santarém (Boomert 1987; Rostain 2006b). It is almost certain that the *muiraquitãs* discovered at the Arauquinoid sites in Guiana originated from Kwatta artisans and diffused along the coast, establishing an effective exchange network. Because these stones were absent on the Surinamese coast, they were obtained through groups along the interior with access to the sources. For example, in the Brownsberg Massif, specialists extracted the stones, worked them, and polished them to make tools, particularly blades for axes (Boomert and Kroonenberg 1977). These products were transported from south to north by the principal riverine routes (the Suriname and the Saramacca) to the coast. Excavations conducted in 2004 by Aad Versteeg and myself on the right bank of the Suriname River revealed the existence of Kwatta settlements at the limit of the coastal plain with the interior. The sites of Rac-a-rac, Surnaukreek, and Km 7.5 were Kwatta occupations that served as exchange localities between the southern and northern communities. Raw material and finished tools were traded with coastal Kwatta populations who could in turn provide smoked fish, crabs, mollusks, turtle eggs, etc., to the inland populations. Artifacts of worked or raw stone, as well as shell and bone pieces, were therefore objects of trade with other Arauquinoid cultures along the coast.

Similar trade was still practiced during the colonial period, at the end of the nineteenth century:

> There existed among the tribes of this . . . district a coarse distribution system for their work; and this not only serves their immediate goal of provisioning all of the tribes with best goods that each can produce, but it also brings together the different tribes and spreads among them ideas and news of general interest. . . . Each tribe had produced goods particular to them; and their members constantly visited other tribes, often hostile, with the goal of exchanging products of their own labor for those produced by the other tribes. (Im Thurn [1883] 1967:270–1).

Curiously, the remarkable remains of this culture have not been studied in detail and no typology has existed for its ceramics until the beginning of the current millennium when I began to study this material, like those of other archaeological cultures in Suriname, from the collections of the Stichting Surinaams Museum. These fine ceramics are tempered with crushed ceramics, but also with *cariapé,* sand, or crushed shell. Many bowls exist, but as in other Arauquinoid cultures, plates are more common, with rims that are frequently decorated with reliefs. The principal decorations are incised lips, wavy cords, incised or punctate cords, and exterior thickenings forming a series of horizontal waves. The typical and very unusual interior rims are formed from several incised thickenings. Small zoomorphs in relief are also found. Furthermore, red paint is common, often applied in bands 2 centimeters wide or more.

Cemeteries in Suriname are located within the settlement sites, often under the occupation level. The skeletons of 38 individuals were discovered at Kwatta-Tingiholo (Khubadux et al. 1991; Tacoma et al. 1991). The skeletons were in extended positions or squatting in the 13 direct burials. Four crania were separated from their trunks, and each was covered with a ceramic vessel. Some bodies presented a slight curving, perhaps due to being deposited in a hammock when buried. Several large, complete ceramic vessels were filled with black earth, while others contained human remains, apparently the remnants of a primary burial. Study of the site's human bone has allowed identification of 21 men, 11 women, and 3 children (and 4 individuals of indeterminate sex and age). The individuals range from 3 to 65 years old, with most between 15 and 35 years old. Only a few individuals were older. Seven crania showed frontal flattening, apparently carried out intentionally on at least three of the men. Seven dental abscesses were noted for six individuals, along with a general increase in the degree of wear on the teeth with age. The lateral incisors were shovel-shaped, a characteristic trait of mongoloid and Amerindian populations (Versteeg et al. 1990). Two men measured 1.69 meters tall, and one woman 1.54 meters. Some pathologies were noted, and for some of the skeletons, study of arrested growth lines (which develop on the long bones during linear growth) indicate temporary conditions of disease or famine. The authors interpret this finding as an indicator of the consumption of manioc rather than maize (Khubadux et al. 1991; Klinken 1991).

The Barbakoeba culture. The Barbakoeba culture occupied the largest territory: between the Cottica River in Suriname and Kourou in French Guiana, an area 200 kilometers long and, on average, 25 kilometers wide. Until recently, this was the least understood Arauquinoid culture of Guiana (Boomert 1993). This is explained by the unexceptional aspect of its material culture and by its location in a region that is difficult of access, leading local archaeologists to seek study sites elsewhere, with more promising cultures. In my last two projects, I was particularly interested in this relatively unknown culture, and thus excavated sites in Suriname and French Guiana. The most significant characteristic of this culture is its specialization in raised-field agriculture. The highest density of structures is found within its territory (i.e., where the last communities of the Arauquinoid tradition occurred). These were the groups that constructed the largest mound complexes and developed the technique to achieve its maximum efficiency.

Coastal villages were located on *cheniers*, spreading parallel to the shoreline, while raised fields were located in the marshes to the south. Like their Kwatta neighbors, they also established outposts leading into the interior. Advanced Barbakoeba settlements are spotted at Petit-Saut on the Lower Sinnamary in French Guiana and at the sites of Surnaukreek and Rac-a-rac on the Lower Suriname in Suriname. These inland settlements likely served as centers where they exchanged crops with forest groups for products that were absent on the coast.

Large cemeteries were juxtaposed with domestic settlements (Rostain 2010b). Excavations at the Sable Blanc necropolis show different groupings of several burials in the same area and a hierarchy of burial types, as well as the quantity of offerings (Van den Bel 2009). However, secondary burials in urns seem to be the dominant type. A large necropolis of Barbakoeba urns is located to the extreme west of French Guiana, under the modern Kali'na villages of Awala and Yalimapo. Urns (generally ovoid in shape and sometimes with small, simple sculpted appliqués) are regularly uncovered during contemporary earth-moving operations (Cornette 1987; Janin 2002; Rostain 1994a; Thomas 2002).

Like their other manufactured products, the ceramics were coarse with simple decoration. These decorations consisted of clumsily executed rows of punctuations or fingerprints, appliqué cords (possibly punctuate), apparent coils, and highly stylized zoomorphic models, particularly with double heads with only eyes that are perforated discs (nose, lips, and ears are absent) (Figure 34). Finally, there are specific bases made from a coil rolled in a spiral, which are also found in Kwatta ceramics. Not surprisingly, simple manos and metates, made from thick, rough-hewn slabs, are often found at these sites. They were likely used to crush the grains and tubers harvested from plants grown on the raised fields. It is easy to imagine that the Barbakoeba groups spent little time on artisanal production or trade. They instead devoted themselves to agricultural specialization.

The dates that have been obtained suggest these sites were inhabited between 990 ± 30 and 825 ± 30 years BP, but it is probable that the chronological range was greater and that the Barbakoeba culture lasts longer.

The Thémire culture. The hybrid aspect of the Thémire culture is its most unusual characteristic. Occupying an area approximately 100 kilometers long and 30 kilometers wide, between the Kourou River and Île de Cayenne, it represents the easternmost representation of the Arauquinoid tradition. At the site of Bois Diable, toward Kourou, it is superimposed over earlier Barbakoeba occupations (Rostain et al. 2009). Indeed, the Thémire culture represents the ultimate manifestation of the Arauquinoid tradition both geographically and chronologically. Most of the dates obtained place it between 510 ± 40 and 260 ± 50 years BP, thus indicating an overlap with the colonial period. Furthermore, artifacts of European origin (glass, earthenware, metal, and even a round ball of lead) have been found associated with indigenous materials at the site of Glycérias at Île de Cayenne (Rostain 1994a). These sites also extended over the *cheniers* or on high riverbanks. Their presence is more tenuous east of Île de Cayenne, but several clues indicate that the Thémire communities may have interacted with inland populations via the Approuague River. I refer to the existence of the exchange of axe blade blanks and preforms originating from southern sources (Rostain 2006b).

Between Macouria and Île de Cayenne, the ceramics lose the Barbakoeba influence and take on a different style (Figure 35). Open forms are common, as are incised rims; parallel vertical, oblique, or spiral incisions; punctuations; appliqué cords; rims with relief; and modeled zoomorphic accents. Painting tends toward white-on-red (or more rarely, the inverse), as well as solid reds. Nevertheless, an important difference in the decoration compared with that of the Arauquinoid tradition of Guiana is noted with the introduction of a foreign style. Spiral motifs and the use of red and white (and sometimes black) paint do not occur in the Arauquinoid corpus. This style is clearly the result of influence from populations on the eastern coast, between Île de Cayenne and the mouth of the Amazon. Certain sherds are clearly imitations of the Late Aristé style or are imported pieces. It is important to recall that some Aristé sites were noted on Île de Cayenne. It is surprising to observe that while

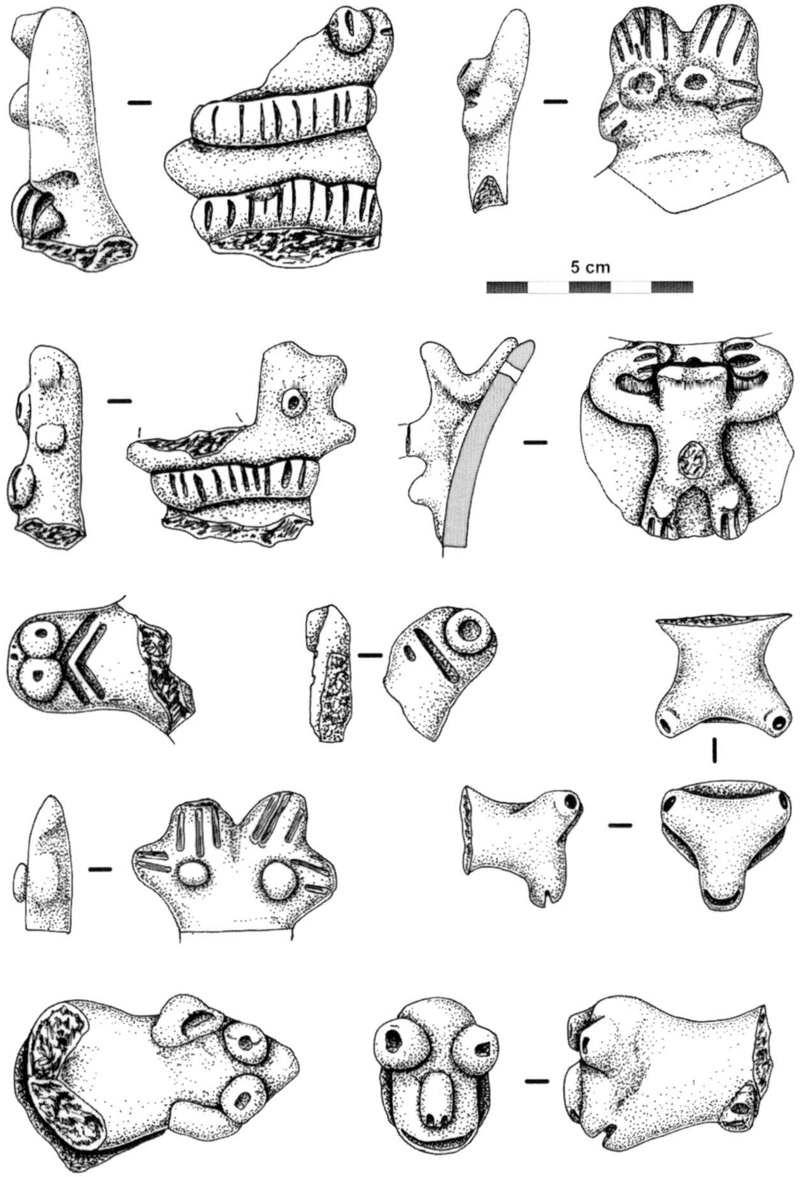

Figure 34. Typical zoomorphic modeled figures of Barbakoeba culture, Sable Blanc site, French Guiana (drawing by Stéphen Rostain).

the Thémire easily integrated the Aristé style, the reverse does not seem to have been the case. It would seem that the Late Aristé communities were less permeable to exterior influences than the Arauquinoid cultures,

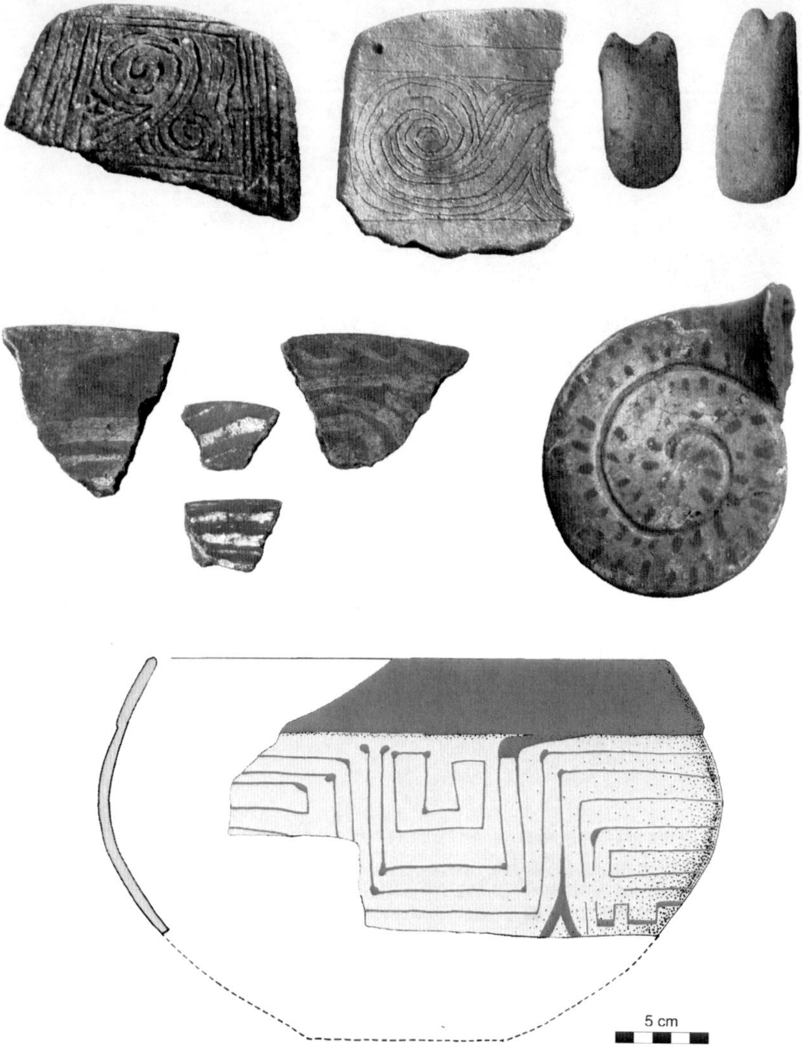

Figure 35. Incised or red-and-white painted pottery of the Thémire culture, French Guiana (photo and drawing by Stéphen Rostain).

which more easily received exogenous elements and coexisted without apparent difficulty with foreign groups.

During the late period preceding European contact, the Île de Cayenne constituted a region that bridged between the cultures tied to the cultural hearth of the Middle Orinoco and those that developed along the Lower Amazon (Rostain 1994c). Although originating from the Arauquinoid sphere, the Thémire culture, because of its frontier position, was subjected to a variety of influences from groups with quite distinct origins, which led to the birth of a new, mixed culture.

The Koriabo Culture: A Guianese Tradition

Cultural upheavals, perhaps linked to climatic changes (Meggers 1994) and manifested principally by the appearance and disappearance of societies, are observed in Guiana beginning around AD 1000 to 1200. However, the most severe shock occurred in the late fifteenth century with the intrusion of European adventurers. The Amerindian world was completely destabilized, and the cultural map underwent profound changes. The arrival of Europeans destroyed the fragile social equilibrium in Guiana.

The period between AD 1000 and 1200 was marked by perturbations of the human geography and cultural composition of Guiana. The Koriabo culture appeared around AD 750 in the interior, and around AD 1100 on the coast. Its place of origin remains uncertain—possibly the Middle Amazon or even the center of Guiana. It is difficult to assign this group to a cultural tradition, as its characteristics are quite unique. It is not clear that an affiliation with the Polychrome tradition is appropriate based simply on the use of multicolored ceramic decoration (Boomert 2004). Koriabo people occupied the banks of major rivers and spread gradually toward the north, moving downstream. Although we can follow the progression of the Koriabo from the interior toward the coast, this culture is sometimes difficult to identify. Many Koriabo sites were established in the same places as earlier sites of other cultures. As the archaeological soils of the Amazon are often very thin and disturbed, it is almost impossible to define the specific characteristics of Koriabo occupation in these mixed sites. They seem to have often reoccupied the villages of other groups. The sites of Rac-a-rac and Surnaukreek in Suriname are good examples that clearly show two successive occupations, the first of which is Kwatta and the subsequent one Koriabo. The dates of 990 ± 40 years BP at Rac-a-rac and 475 ± 40 years BP at Surnaukreek could be attributed to either group. The Koriabo culture continued in certain areas for a long time during the colonial period. Its presence was still confirmed during the second half of the eighteenth century on the Approuague in French Guiana (Rostain 1994a), while farther west, some groups of the same culture migrated to the Lesser Antilles, where they persisted a bit longer (Boomert 1986, 1987).

The Koriabo culture is nevertheless unique because it represents the only truly Guianese culture, perhaps of local origin, that is not found outside of this region. It appears almost clearly and simultaneously throughout Guiana, from the Oyapock to the Orinoco Rivers, and from the center of Guiana to the coast. Furthermore, the truly remarkable aspect of the Koriabo culture is the stylistic uniformity of the shapes and decoration of its ceramics throughout this immense region. The type of clay used obviously varies from one region to another depending on the local geology, but the shapes of vessels are repeated everywhere and the very particular incised and appliqué decorations are found at all of its sites, even over distances of hundreds of kilometers.

Koriabo ceramics are very distinctive for their shapes as well as for their decoration. The temper varies considerably from one region to another: fairly fine sand, crushed quartz, mica, crushed shell, *cariapé*, etc. The two most standard vessel forms are globular jars (often with four lateral bulges on the body) with a vertical neck and outflaring rim, as well as bowls with concave, everted walls and multilobed rims. The thick rims with flat lips, with wide incised lines, are also characteristic. The decoration combines fine incised lines and very wide, shallow incisions made with a serrated stick, cords that are sometimes incised, annular disk appliqués, and white, black, and red paintings. The motifs are scrolls and spirals. Sculpted relief is sometimes very elaborate, depicting hybrid human/animal figures. Felines (jaguars or ocelots) dominate the realistic depictions of animals.

The Koriabo maintained special relations with the Aristé (Rostain 1994a). The Koriabo descended far along the Oyapock River, close to its mouth (i.e., on the border of Aristé territory), but they did not settle the core of this zone, in the hills of Oyapock Bay. Indeed, only one Koriabo jar, decorated with four faces, was found placed in a small rocky cavity, suggesting a votive offering. Beyond this example, not a single Koriabo sherd has been discovered in Aristé sites (Rostain 1994a). In contrast, Aristé influence is reflected at times in Koriabo pottery. It is perhaps this influence, which Arie Boomert detected, that indicates more than a "genetic" link between Koriabo culture and the Polychrome tradition (Boomert 2004). A remarkable jar was collected from a tributary in western Suriname, a great distance from the Aristé sphere of influence. Nevertheless, its form, temper, painted decoration, and skill of its execution are all characteristics of the Aristé style. Only the animal represented, an ocelot (recognized by the black mask around its eyes), does not belong to the iconographic corpus of the Aristé, but rather to that of the Koriabo. Does this represent a vase (perhaps funerary) created by an Aristé potter that was transported long distance to a Koriabo partner? Was this piece created by an Aristé woman who was inspired by a Koriabo warrior? Is it evidence of a clan association between the two cultures? Although there are many possible answers to these questions, this piece indicates a specific link between Aristé and Koriabo.

The Koriabo invasion of the Guianese coast was not, however, complete. The Hertenrits territory of Suriname and the Aristé space of Amapá seem to have been little touched by this culture. In contrast, the remainder of the coast was greatly affected by Koriabo influence. Whether on the coast of Guyana or the coast of Suriname and French Guiana, all the Arauquinoid sites present evidence of a Koriabo presence in their ceramics. Beginning in AD 1200, the coastal populations of Guiana evolved under cultural pressures that originated from the interior. The Hertenrits culture apparently progressively disappeared from AD 1250, like the other Arauquinoid cultures that gradually extinguished one by one from west to east. The Thémire culture was the only one that appears to have survived until the moment of European contact.

Although little affected by the Koriabo expansion, the Aristé people of eastern Guiana experienced a change provoked by the growth of the Polychrome tradition of the Lower Amazon. Incised techniques gave way to painted ceramic decoration, and motifs became refined. The Aristé groups survived into the first centuries of the colonization.

The domination of the coastal Arauquinoid populations seems to crumble at the beginning of the second millennium AD. They gradually disappeared to the west (the Kwatta and Hertenrits cultures), while to the east they took on cultural traits of groups that came from the interior (the Barbakoeba and Thémire cultures). During the same period, the Aristé groups from eastern Guiana underwent profound changes. The coastal communities of Guiana thus experienced phenomena of decline or change at the beginning of the second millennium AD. In contrast, the arrival of Europeans in the fifteenth and sixteenth centuries dealt a fatal blow, provoking the complete destabilization of the indigenous world.

Then Came the White Men

1492: the indigenous discover the Europeans, which leads to the rapid disappearance of a world due to the microbial shock resulting from contact with the new arrivals. Meanwhile, the continent gradually fell under the yoke of the conquerors. The Conquest, which began in the early sixteenth century, unleashed considerable chaos in the indigenous world, which had been apparently already changing over the previous four centuries. Imported disease decimated groups and destabilized the indigenous populations. Interethnic warfare took a new turn, often exacerbated and manipulated by the European conquerors. Fleeing from the troubles created by the new arrivals, entire indigenous populations migrated across Guiana. Cultural divisions and crossings increased, resulting in a total restructuring of indigenous societies (Rostain 1994a; Whitehead 1993, 1994, 1996). Indigenous Guiana rapidly adopted a new face that had no relation to that visible prior to the Conquest.

Nevertheless, the impact of the meeting of these two worlds in Guiana must be put into perspective, at least when compared with the toll it took on the Amazon and the Orinoco. Compared with these two regions, Guiana suffered proportionally less from Western abuses. It is legitimate to wonder about the fate of the archaeological cultures present in Guiana at that time. While the Arauquinoid populations to the west, in Suriname, had been experiencing a decline at this point for three centuries, those to the east, in French Guiana, were flourishing. Moreover, the Koriabo and the Aristé communities were at their peak.

Although the Europeans frequently visited their territory since the beginning of the Conquest, the Aristé were initially less affected by this contact than other groups. Control of this coast represented a less critical priority for the European nations, who fought more violently for control of the two major rivers, where indigenous groups banned together before the new arrivals; through the direct and continuous contact with these foreigners, they were affected much more violently by disease epidemics. Moreover, the populations dropped at a dizzying rate in a very short period of time, explaining why the Caviana, Aruã, Mazagão, and Maracá cultures disappeared shortly after the Conquest.

The arrival of the Europeans encouraged indigenous migrations and the flight of groups attempting to escape persecution. The archives indicate the abandonment of the Lesser Antilles, the Orinoco Delta, and the Lower Amazon. Some of the displaced tribes came to settle in Amapá, apparently a less turbulent location. They mixed with the Aristé, disturbing their stability. New combinations of ethnicities and stylistic mixing began, leading to the emergence of reconstructed hybrid communities, among which are the modern Palikur of northern Amapá.

The presence of Dutch faïence and glass beads in the Terminal Aristé necropolis in the hills of the Lower Oyapock permit the dating of the final manifestations of this culture to around AD 1750. Nevertheless, the decoration and form of the final Aristé funerary ceramics from the colonial period already clearly reflected the appearance of new influences and fusion with foreign styles.

The historic continuation of the Aristé culture justifies a comparative study with the dominant local indigenous groups since the colonial period, principally the Palikur. The analysis is based on the comparison of archaeological Aristé, Koriabo, and Arauquinoid ceramics with modern Palikur vessels for their form, decoration, and techniques. While conducting an ethnographic study among the Palikur, I showed some potters photographs and drawings of archaeological vases, and asked them if they recognized them (Rostain 1992). I also collected data from the colonial archives and the Palikur oral tradition. This work allowed me to reconstruct the possible evolution of the Aristé culture to that of the modern Palikur. Stylistic study of Amapá ceramics shows a gradual change that began with European contact (Rostain 1994a). This gradual change is particularly evident in the funerary pieces. Late Aristé funerary vessels (AD 1100–1600) were decorated with very fine sinuous and spiral polychrome motifs in red, black, brown, and white. The urns display a human silhouette, stylized with appliqués representing the face, arms, and legs. The Palikur do not recognize this style. The material from Terminal Aristé cemeteries (AD 1600–1750) that contains European artifacts is made up of square, oval, or globular urns, objects with completely new shapes, and decoration that is principally black-on-white, generally limited to the neck. The Palikur recognize this style, which recalls that of the Mazagão and Arauquinoid cultures; comparable motifs can be found in their contemporary art.

This change in ceramics indicates the emergence of exogenous styles in Aristé pottery, which resulted from the arrival of refugees from the Lower Amazon and the western coast of Guiana shortly after the Conquest. The form and the decoration of the vessels deviate gradually from the key characteristics of the Late Aristé and resemble much more closely traits of the Mazagão and the Arauquinoid. This period of stylistic transition corresponds to the Terminal Aristé, and led to the creation of a composite culture, whose modern representatives are the Palikur. The colonial fusion of Aristé, Mazagão, and Arauquinoid styles thus constitutes the genesis of Palikur art. At the beginning of the twentieth century, and probably just before, it possessed all its modern attributes (Nimuendaju [1926] 2008), and no major indigenous migration or major exterior influence has altered it since that period. This evolution was accompanied by the progressive abandonment of ancient funerary customs of cremation, dominant in Late Aristé culture, and the adoption of secondary burials following defleshing, which is more typical of the Mazagão and Arauquinoid cultures. This historic schema agrees with the origin stories of the Palikur, both in the oral tradition and in written archives (Grenand and Grenand 1987).

Two cultural complexes therefore appear to have affected the Aristé after the Conquest. The first, associated with the Arauquinoid tradition, came from western Guiana. The second originated in the Lower Amazon and appears associated with the Mazagão culture. The archives report that indigenous peoples arrived in eastern Guiana as they fled the European hunt for slaves, which was particularly intense in the Lower Orinoco. Some of these refugees of Arauquinoid culture likely brought with them the raised-field technique that was used by the Palikur in the eighteenth century, when they became too numerous to support themselves using the slash-and-burn technique (Grenand 1981; Rostain 1991).

The immigration of populations that originated at the mouth of the Amazon to Amapá is mentioned in both archival documents and Palikur oral tradition. The latter speaks of two Palikur clans that arrived from southern of Amapá and introduced the modern Palikur language, manioc cultivation, and manioc graters with metal teeth (Grenand and Grenand 1987). Some of these clans previously lived on the banks of the

Maracá River, the cultural hearth of the Mazagão culture. They were therefore likely associated with this culture and brought several of its characteristics with them.

Whereas the modern Palikur are heirs to the Aristé communities, and are connected as well to the Mazagão and the Arauquinoid, other groups were involved with their emergence. During the first centuries of European contact, the northern part of Amapá was home to 18 Palikur clans and 23 other ethnic groups, the most important being the Aruã, the Arikare, and the Maraon (Grenand and Grenand 1987). These different groups were no doubt affiliated with the Aristé culture that still dominated the region. The pattern could be compared with the modern circumstances on the Upper Xingu, where 16 ethnic groups of the Gê, Aruak, Tupi, and Karib linguistic families share the same material culture, with each group specializing in the production of certain goods. For example, one group produces all ceramic goods, which are then traded to their neighbors. Thus, an archaeologist who excavates sites in this region would find a unique style that he might consider to represent a single culture. However, an archaeological culture can reflect the existence of several ethnic groups linked by a common material culture.

The study of the colonial archives gives the impression of a confederation in Amapá that integrated various ethnic groups within a common ceremonial and exchange system, and which united occasionally under the authority of a war chief to carry out military actions (Grenand and Grenand 1987). This minor military confederation, coupled with the existence of a language of trade and ceremony as well as the importance of festivals between clans and ethnic groups, are recalled with nostalgia by the Palikur and are signs of an ancient unity that was fueled by intense bartering. A comparable situation likely existed for the Aristé during the pre-Columbian period. Similarities in material culture, precise territorial boundaries, and the rarity of foreign artifacts are indicators of this culture's strong homogeneity. Several communities shared a common material culture and lived in a dense trade network.

The limited number of urns in each Aristé necropolis and presence of greater wealth in some suggest that this burial practice was reserved for the elite, clan chiefs, or shamans (Rostain 1994a). Furthermore, the abundance of separate cemeteries, each with few urns, and the variety of ceramic decoration perhaps indicates ethnic divisions. All of these characteristics may be specific manifestations of the identity of individual groups within a homogeneous cultural complex. The existence of different cemeteries, which today represent the clan divisions among the Palikur, no doubt reflects the ethnic diversity of the ancient pantribal confederation.

And finally, what of the Kali'na, who have dominated western Guiana from the Île de Cayenne since at least the Conquest? This region corresponds exactly to the Arauquinoid territory. Nonetheless, few apparent resemblances are detected between the two cultural complexes. Kali'na ceramics, collected by museums since the eighteenth century, do not display any affinities with those of the Arauquinoid. They are much more similar to Aristé art based on their scroll motifs and polychrome decoration. I suspect that this is the result of a phenomenon similar to that which produced modern Palikur ceramics. Aristé groups migrated toward the west, bringing with them their knowledge and their style, which had already influenced Thémire and Koriabo societies. They mixed with the coastal Guianese populations and easily imposed upon them their own iconography, which quickly absorbed the local styles. This mixture, dominated by the Aristé, and with less marked contributions from the Arauquinoid and Koriabo, quickly led to the imagery of the modern Kali'na.

Although Guiana was colonized slowly, the archaeological cultures were extinguished quickly and the indigenous populations plummeted. Modern indigenous peoples no longer have much in common with their pre-Columbian ancestors.

A Tale of Tapirs . . .

Reconstructing the pre-Columbian occupation of Guiana in five principal stages allows us to observe the key transformations that these societies have undergone.

More than 10,000 years ago, the first inhabitants roamed the savannas of central Guiana to hunt large animals. It is likely that their presence then began to affect the environment, as they encouraged certain vegetal associations (Cooke and Piperno 1996; Politis 2001), and they may have enlarged the savannas through burning.

The fisher-collectors that followed, 7,000 years ago, did not provoke major environmental change through their consumption of marine mollusks. Nevertheless, the enormous shell middens upon which they established villages indicate the beginning of sedentism in Guiana. Furthermore, after initially probably using plants as containers (Roosevelt et al. 1991), these groups advanced technologically, developing ceramic technologies.

Around 2000 BC, groups began to construct villages, produce tools and objects, and cultivate slash-and-burn fields in the forest. The cultivators crossed their territory on foot or with boats on river routes. As time went on, the environment became more intensively exploited and transformed. Effects of these anthropogenic impacts are still felt in the rainforest today. It is clear that the period between 7,000 and 4,000 years BP was a time of significant cultural developments that included the invention of ceramics, sedentism, and plant domestication. During the first few centuries of the Common Era, societies in favorable areas increased in complexity and population size, increasing their production and specializations.

Guiana's Neolithic revolution began around AD 300 and was in full force by AD 600. Riverbanks and the marine coast were inhabited by populations that controlled and transformed their environment. The sociopolitical structure of these societies evolved with the appearance of hierarchical organization, task specialization, trade systems, and shared ceremonial practices. On the coast, cultural spaces were clearly defined, usually demarcated by the major rivers. Populations grew, leading to an increasingly complex distribution of territories and more precise cultural identities.

Two hundred years ago, the Dutch philosopher Cornelius de Pauw (1770:iv) wrote: "There is almost nothing left of ancient America except the sky, the earth, and the memory of its terrible misfortunes." But although the pre-Columbian history of America's indigenous peoples may not be written in books, it is nevertheless present in its soils, stones, trees, and rivers; it is everywhere in the Amazon. And it is the task of archaeologists to rediscover and restore this rich past.

One of the unique characteristics of Amazonian archaeology, which distinguishes it from that of other regions of the Americas, is the importance and density of the phenomena observed. Indeed, the principal cultural traditions of the Amazon exhibit remarkable chronological depth and geographic extents. It is not unusual for a tradition to persist for centuries (even one or two millennia). Furthermore, these traditions occupied immense territories, such as the example of Arauquinoid groups, who dominated the Orinoco Basin, the coast of Guiana, and the Middle Amazon for more than a millennium. Certain similarities are even observable in the contemporaneous ceramic styles of the Llanos of Mojos in Bolivia and the Lesser Antilles. The Polychrome tradition is also recognized during the first centuries of the Common Era all along the Amazon Basin, and persists even today with the Shipibo-Conibo of Peru, the Quechuas of Ecuador, or the Kali'na of Guiana. This characteristic far-reaching geochronology is a key factor in understanding the unique evolution of the early settlements of the Amazon.

Even more extraordinary are the parallels observed between the traditions of the Amazon Valley and those of the Orinoco Basin. It seems that the cultural histories of each of the two rivers ran closely in parallel. Betty Meggers and Clifford Evans (1961), followed later by Peter Hilbert (1968), present a chronocultural framework for the Amazon that is quite similar to that proposed independently by Irving Rouse and José Cruxent (1958–59) for the Orinoco. As a result, the Hachured-Zoned tradition of the Amazon resembles the Saladoid tradition of the Orinoco in many aspects. The Amazonian Incised-Rim tradition that follows is analogous to the Barrancoid tradition of Venezuela. The Incised-and-Punctate tradition of the Amazon shares many identical characteristics with the northern Arauquinoid. Only the Polychrome tradition appears totally unique and native to the Amazon Basin.

I have noted the striking similarities between the Arauquinoid ceramics of the Guiana coast and the Santarém style, but similarities can also be found with the ceramics of other builders of raised fields in the Llanos of Mojos in Bolivia. Analogies can also be found between the pottery of the Arauquinoid of the Guianese coast and the Orinoco, and that of some sites in the Dominican Republic. Based on the geographic regions, the archaeological data, and the cultural characteristics, it is possible to distinguish at least three large complexes within the Arauquinoid tradition: Orinoco Apure, Guiana, and the Lower Amazon.

Despite these obvious cultural and chronological parallels, it is difficult to imagine that cultural hearths on one river could have diffused so rapidly to the other basin. It is more reasonable to assume simultaneous innovations occurred in the two most dynamic cultural centers of the tropical lowlands.

For 50 years, Amazonian archaeologists have fiercely debated the polemic regarding the potential for cultural development in the world's largest rainforest and the pre-Columbian state of Amazonian societies. According to Meggers (1996), it is impossible for a complex society to have developed in such a difficult environment. Instead, the few complex societies that have been observed archaeologically, such as the Marajoara, must have originated elsewhere (e.g., the Andes). The principal opponent to this view, Lathrap (1970), argues that the Middle Amazon functioned as a heartland for cultural innovation from which spread various populations throughout the region. Work carried out by various archaeologists since Roosevelt's (1980) excavations tend to support this latter model, although they indicate that there were in fact multiple cultural heartlands (rather than a single source) distributed throughout the Amazon.

Finally, the principal fault of some authors may be to assume that demographic growth in the Amazon resembled a regularly rising curve, without considering the heterogeneity of the landscape and its variability through time. Humans did not equally exploit all of the tropical biotopes; they found the open areas of the savannas and the *várzea* much more appealing. The agricultural and diverse resources also vary to great extent between the different natural habitats (Bush and Silman 2007). Although large and dense societies may have developed on the alluvial plains, the interior savannas of Guiana do not seem to have experienced comparable demographic growth. Whether considering Formative period fisher-gatherers or more recent farmers, areas of *várzea* were clearly more favorable for sedentism than other regions of the Amazon. In contrast, since the time of the initial people of the Amazon, the interior savannas of Guiana have been more often used by nomadic and seminomadic peoples. Even today, indigenous people often exercise a high degree of mobility in this region, where a dense communication network is maintained by a system of trails.

The Amazon offers a great variety of natural environments. These landscapes have also greatly evolved over the course of the Pleistocene and the Holocene. It is certain that this mosaic of environments corresponded to great cultural diversity in the pre-Columbian period, sometimes observable in the archaeological record. Perhaps even more than the Andean highlands, the tropical lowlands once supported a wide range of diverse indigenous communities, which were later subject to social homogenization due to the impact of the European Conquest.

Chapter 4

A Natural Garden or a Domesticated Forest?

If we omit the extreme Cartesianism of the colonial polders of Suriname and Guyana, modern coastal Guiana observed from above could resemble the ancient Roman army. We see centuries, maniples, and cohorts of mounds tactically arranged in square or tortoise formation. They are found along thalwegs or sandbars, arranged in Praetorian squadrons surrounded by marshes or extending across the savanna in Hoplite phalanges. Whatever their form, these transformations have vanquished the chaos of the tropical coast. The carefree liberty of nature has surrendered to indigenous ingenuity. The blade of grass has been conquered, crushed, and annihilated by humans, who have put in its place the vegetation of their choosing. This contrasts with the traditional French farmstead, characterized by centuries of successive buying and selling as well as conflicts with neighbors, altogether a turbulent legacy. Guiana's patchwork landscape reflects the harmonious and flourishing product of communal labor. A mighty general has subdued the barbarous equatorial nature. Which tribunes and centurions organized this structured and disciplined legion?

There are two principal methods to cultivate crops in wetland environments. The first is polderization, which consists of draining the plots by using a checkerboard of canals. Europeans, particularly the Dutch, have been using this method on the coast of Guiana since the seventeenth century. Hundreds of hectares of abandoned polders are visible east of Paramaribo, Suriname, while others are still used on the east coast of Guiana. The second technique is the use of raised fields in marshes, which indigenous peoples employed along the coast of Guiana for close to a millennium before the arrival of Westerners. It is the inverse of the polders because instead of digging canals to drain the fields, the mounds are built to rise above the natural flood level. Apart from drainage, another potential advantage is that the organic material that is present on the surface of the marsh can be concentrated on the mounds. Because these structures do not always protect crops from flooding, canals and ditches permit an additional level of control on the water level.

Raised Fields of Guiana

Shapes and Dimensions

Although Latin American raised fields may differ in their shape, dimensions, and distributions, most are elongated and of extremely variable size (Denevan 1970). However, the raised fields of Guiana present an anomaly because most of these structures are round or square. Only a few examples of this morphology are found elsewhere, such as occasionally in Bolivia, for example (Denevan 2001). The great variability among savannas does not permit the construction of strict typologies. Nevertheless, the diversity of agricultural structures does not appear random, but is rather the result of careful strategies adapted to the local hydrology, chronology of construction, and the available technology. The form and size of raised fields would be thus influenced by cultural and technological criteria, which thereby justify the development of a general classification scheme. Another important variable is the proportion of cultivable area across the landscape: the less cultivable land, the more important the seasonally flooded zone is for fishing or as a source of sediment that can be used as fertilizer. I propose four major types of raised fields, based principally on size, and to a lesser extent on morphology: elongated raised fields, large mounds, medium mounds, and small mounds. This typology does not include structures that are obviously colonial or modern, such as Creole agricultural beds and the small, modern raised fields.

Elongated raised fields vary in appearance from place to place, with dominant types apparent in different countries (Figure 36). In French Guiana and eastern Suriname, they can range from 1 to 7 meters wide, 5 to 30 meters long (although they are rarely longer than 10 meters), and 0.5 to 1 meters high. They are generally rectangular, although they are occasionally oval. Although sometimes located in flooded depressions, they are more often located along the interface between flooded zones and sandbars. Their distribution, whether parallel or perpendicular to the slope, depends on their elevation. Raised fields at the foot of Quaternary sandbars follow the slope to facilitate drainage. Those located in the non–seasonally flooded zone on top of the sandbars are arranged perpendicular to the slope to optimize water retention. In the year-round seasonally flooded zones, they are organized parallel to one another, forming a checkerboard that facilitates irrigation.

In western Suriname, one finds regular and irregular complexes (Boomert 1976). The former exist in various forms: round, oval, square, and rectangular. They measure 3 to 4 meters wide and 4 to 30 meters long, and the canals between the mounds measure 0.5 to 1 meter wide. They are arranged around habitation mounds in the marshes of the recent coastal plain or close to the sand ridges to the east. Irregular complexes

Figure 36. Elongated raised fields of Guiana. Top: Piliwa complex, Lower Mana, French Guiana (photo by Stéphen Rostain). Bottom: Berbice River, Guyana (photo by George Simon).

are rounded or ovoid rectangles measuring 4 to 5 meters wide and anywhere from 8 to 140 meters long, with 30 meters being the average. The canals between the mounds are narrower, 1 to 2 meters wide, but erosion has partially destroyed them. They are distributed in compact groups of 2 to 15 mounds in marshy areas of the recent coastal plain or below the sand ridges. Raised fields are generally smaller and shorter on the sand ridges than on the coastal plain. On the western part of the coast of Suriname, raised fields can be differentiated chronologically based on their morphology and their association with Barrancoid or Arauquinoid habitation mounds (Boomert 1980b). Regular (quadrangular) raised fields are associated with Wageningen-1 and -2, Buckleburg-1 and -2, and de Burnside habitation mounds. Irregular raised fields are generally near Hertenrits and Wageningen-1 mounds, indicating an Arauquinoid association (AD 650–1200).

Several complexes of elongated raised fields in eastern Guyana are visible in aerial view. Generally, these constructions are poorly documented (Boomert 1978), but those observed are most often distributed perpendicular to a river. Close to Fort Nassau, along the Berbice River, George Simon (n.d., cited in Plew 2005) recorded 787 rectangular raised fields that measure 1.7 to 6.9 meters wide, 5 to 8.2 meters long, and 0.5 to 1.7 meters high. It is likely that other such complexes, associated with habitation mounds, exist along both sides of the Canje and Berbice Rivers.

Large mounds vary in diameter from 2 to 6 meters, and in height from 0.5 to 1 meter (only rarely 2 meters high) (Figure 37). They are generally rounded near Kourou, and quadrangular (although eroded) around the Maillard savanna to the east. Some are oval, rectangular, or even polygonal, such as in the Iracoubo sector. The largest mounds are generally located in the most inundated areas, their size diminishing progressively with their distance from the seasonally flooded zone. They tend to be arranged in a tight gridded pattern, except in Guyana, where the large round mounds are widely spaced. They are clearly visible in aerial photographs.

Medium mounds are round or square (more rarely rectangular). They measure 1.5 to 2 meters in diameter and 30 to 50 centimeters high. They are organized either in groups or distributed in a checkerboard pattern. They form tight groups when located in forested settings and are difficult to detect from aerial photographs alone. These mounds are often very eroded, but likely are identical to the intact mounds that are located in marshy zones.

Small mounds are almost always round, and are only rarely square. They measure 30 to 150 centimeters in diameter and 20 to 30 centimeters high. They are found on the edges of sandbars or on savannas that are slightly seasonally flooded, but are completely dry in August. They are barely visible on aerial photographs at a 1:10,000 scale and are detected though pedestrian survey.

Large and medium mounds are the most common types; small mounds are the most rare (or perhaps only the most difficult to detect). Elongated

Figure 37. Large, round, raised fields of the Diamant complex, west of Kourou, French Guiana (photo by Stéphen Rostain).

raised fields represent an adaptation to specific geomorphological conditions. In most cases, the complexes are made up of an assortment of different types of mounds. This mixing is not accidental. Stereoscopic interpretation of aerial views suggests an intricate engineering at work that addresses multiple factors and is well adapted to local conditions.

In 1991, botanical transects were recorded in the Kourou (K-IV) and Matiti (T-1) complexes of Malmaison in the Maillard savanna (Rostain

1994a). Vegetation communities were characterized by 100 percent coverage of *Cyperus articulatus* and *Melochia villosa,* typical of a high, humid savanna that transitions to rainforest. The humid matrix was characterized by *Fuirena umbellata* and *Fimbristyllis cymosa.* Humid savanna species (*Heliconia psittacorum* and *Sauvagesia erecta*) are found on mounds, associated with secondary savanna taxa (*Paspalum,* various Papilionaceae, *Ludwigia nervosa*), and with shrubby thickets (*Dioclea,* various Sapindaceae).

Raised-field Complexes

During the 1980s, I categorized the raised fields found between the Île de Cayenne and the Karouabo into seven groups, all situated in homogeneous savannas (Rostain 1994a). From west to east, they are Sinnamary (S), Malmanoury (Y), Karouabo (A), Diamant (D), Kourou (K), Matiti (T), and Maillard (M). Within each group, the raised-field complexes are designated by a code (e.g., in western Kourou: K-I, K-II, K-III, etc.). In the last 10 years, I completed mapping of the coast of Guiana, adding the mounds between the Karouabo and the Maroni. Each group is named for the closed savanna upon which it is located. From the Malmanoury to the Mana River, there are the Renner Savanna, Combi, Brigandin Savanna, Corossony North, Corossony Savanna, Trou Poisson Savanna, Henry Savanna, Counamama Savanna, Fiévée, Yanou Savanna, Mathieu Savanna, Éloge Savanna, Roches Blanches Savanna, Mamaribo Savanna, Flèche Savanna, Petit Macoua Savanna, Macoua Savanna, Grand Macoua Savanna, Organabo Savanna, Terre Rouge, Oratoire, Mana-Awala, Piliwa Est, and Piliwa. The map can be broken down into four main sectors: Macouria, Kourou, Iracoubo, and Awala (Figures 38–41).

Multiple survey methods (stereoscopic interpretation of aerial photographs and aerial reconnaissance combined with pedestrian survey) were necessary to produce a precise map of pre-Columbian structures on a 200-kilometers stretch of coast in French Guiana due to the low visibility caused by forest growth. In the Matiti savanna to the east of Kourou, a checkerboard of mounds was detected after the area was cut and burned to clear an agricultural field. Groundtruthing confirmed the existence of the raised fields, but it was nevertheless difficult to distinguish the alignments because the trees had partially destroyed the mounds. The phenomenon of partial destruction of mounds was present in several complexes that were difficult to identify from an aerial view. Canals and causeways are present as well, although in several cases it is difficult to verify that these are of pre-Columbian origin. Although the canals are often associated with raised fields, there are cases where the superposition of construction from different time periods may have occurred.

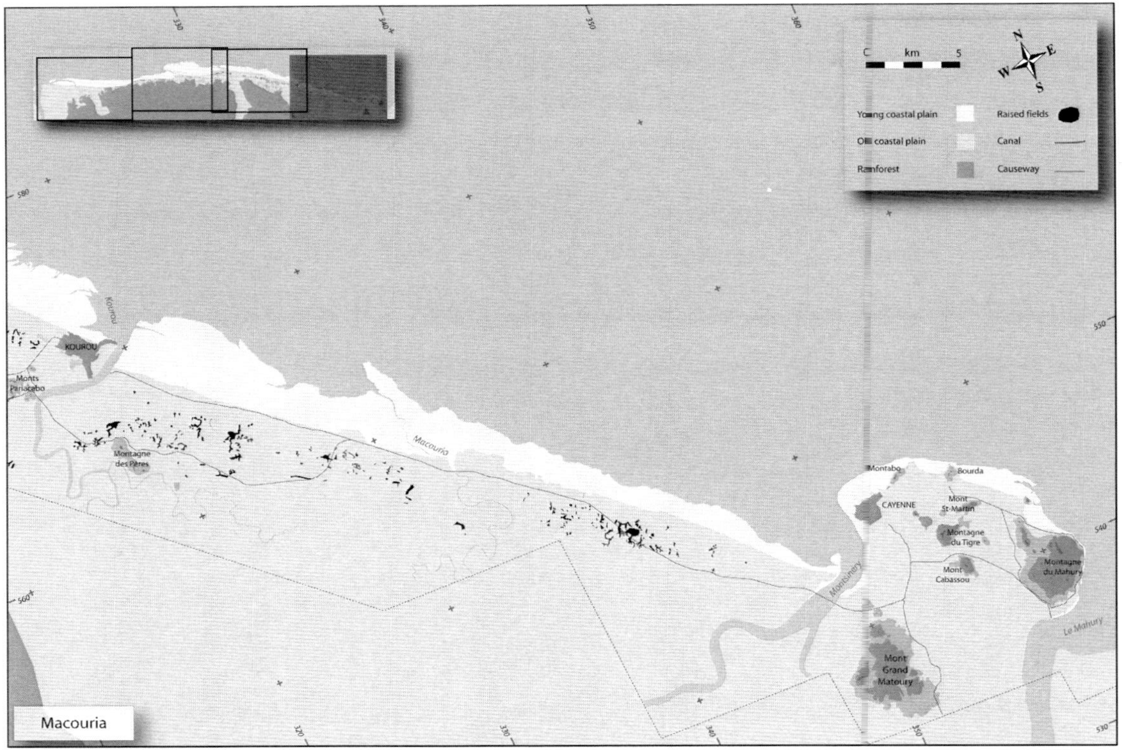

Figure 38. Raised-field complexes of French Guiana, Macouria sector. Coordinates from Centre Spatial Guyanais (CSG) geodesic system (drawing by Stéphen Rostain and Laure Déodat).

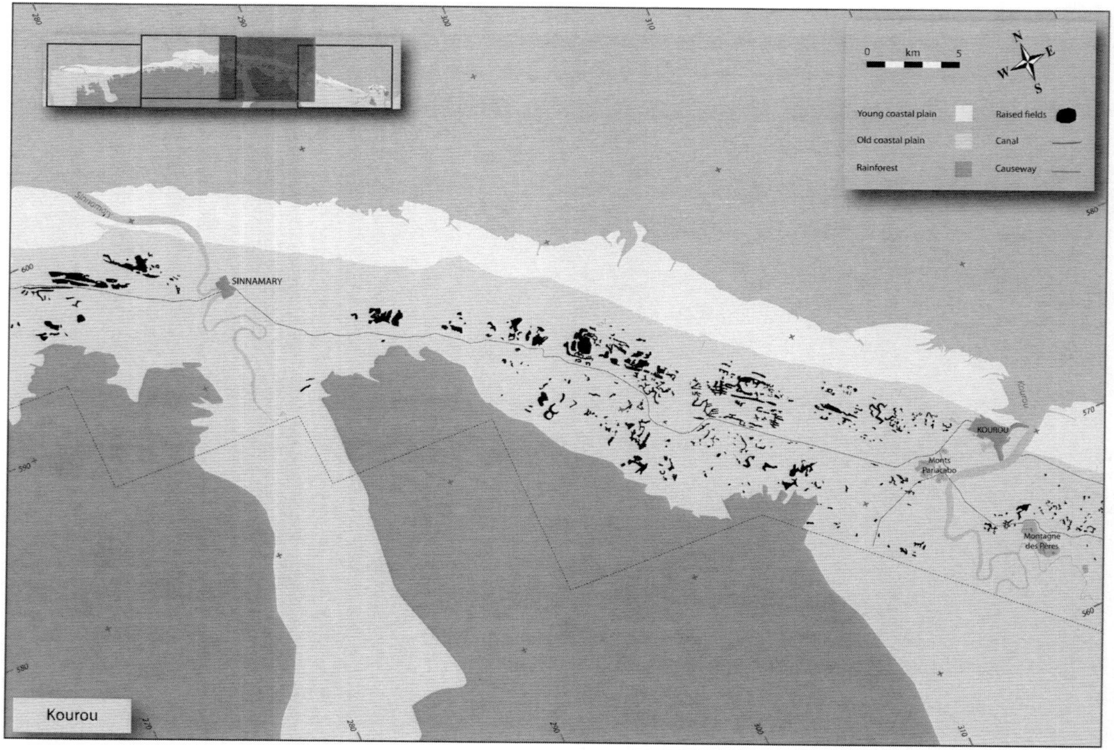

Figure 39. Raised-field complexes of French Guiana, Kourou sector. Coordinates from CSG geodesic system (drawing by Stéphen Rostain and Laure Déodat).

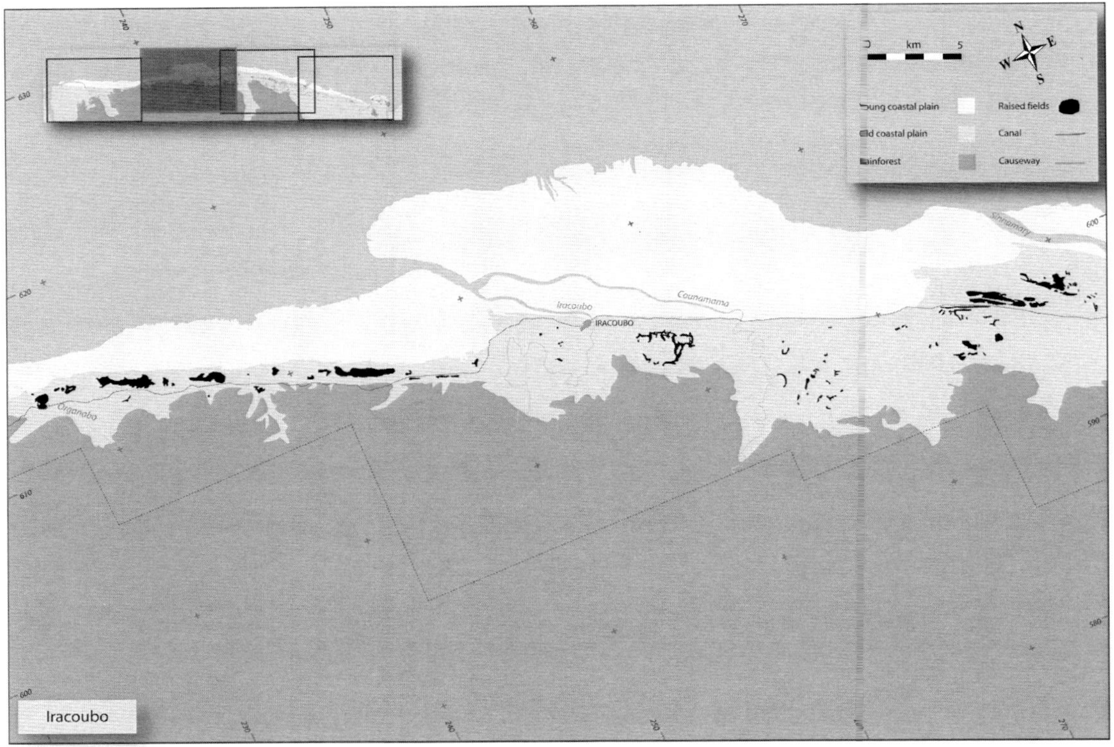

Figure 40. Raised-field complexes of French Guiana, Iracoubo sector. Coordinates from CSG geodesic system (drawing by Stéphen Rostain and Laure Déodat).

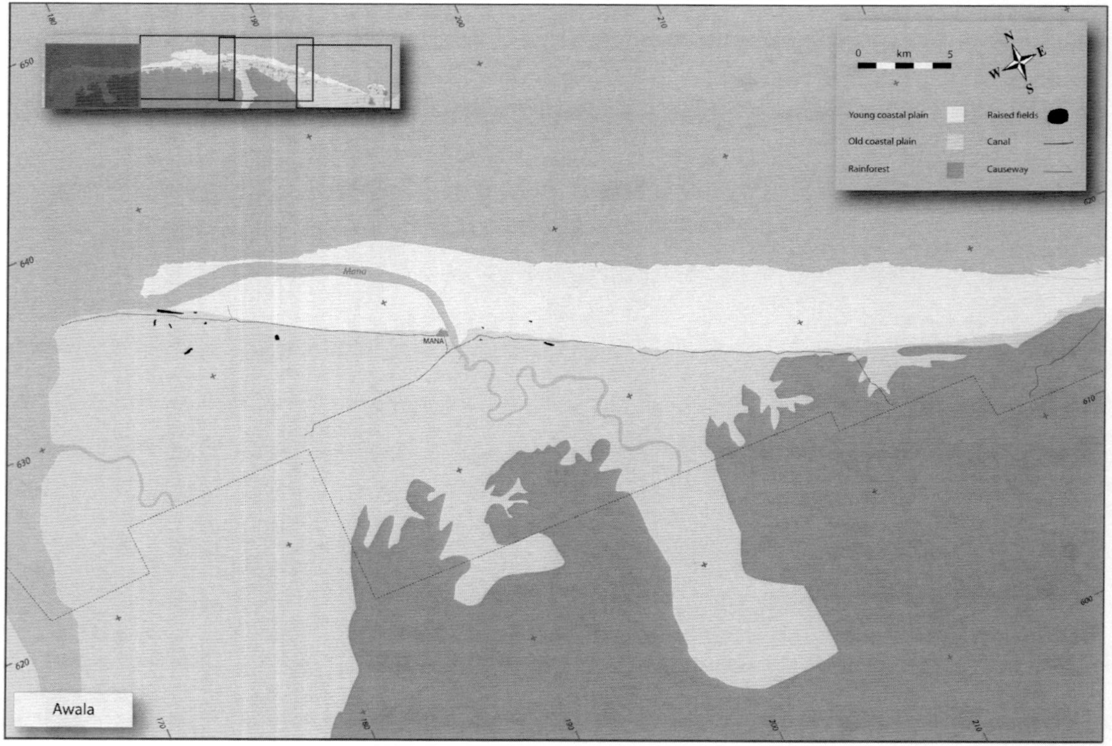

Figure 41. Raised-field complexes of French Guiana, Awala sector. Coordinates from CSG geodesic system (drawing by Stéphen Rostain and Laure Déodat).

Frans Bubberman carried out similar mapping work on a larger scale from 2006 to 2008 along the coast of Suriname (Figures 42–44). Agricultural construction is less frequently recorded over a great deal of this coast because it is hidden by colonial polders. However, it is likely that raised fields were equally common as in French Guiana before European contact. Combining the ensemble of mapping data available across these regions, it is possible to locate the raised fields of Guiana from Cayenne (French Guiana) to the Corantijin River (Suriname), a territory that covers 586 kilometers. The only portion that is currently without data is eastern Guyana, consisting of at least 40 kilometers bordered by the Berbice River.

Mapping of raised fields in French Guiana indicates 662 complexes that cover 2,856 hectares of human-modified savannas. These figures include only the ecofacts that are still visible; it is likely that entire complexes have been destroyed by modern land use. The construction of modern polders in Mana destroyed previous pre-Columbian structures; only vestiges of the mounds are now visible. Elsewhere, highway and other construction projects have also buried mounds. Human abandonment of the savannas has also allowed recolonization by the forest, hiding the remnants of agricultural features. A comparison of aerial photographs of an abandoned savanna to the south of Sinnamary that were taken 40 years apart suggests the forest boundary advanced 13 meters (Jean-François Turenne, personal communication, 2009). In view of these multiple destructive factors, it is reasonable to assume that a large portion of the pre-Columbian constructions of French Guiana has disappeared, and that the extent of territory transformed by indigenous peoples was greater prior to the Conquest.

As mentioned in this chapter, the form and organization of raised-field complexes varies through time and space. The variability of raised fields in Guiana is tied to cultural, chronological, and technological differences. Differences in local hydrological and edaphic factors also play a role.

In French Guiana and Suriname, raised fields are not distributed along rivers, unlike the elongated raised fields of Guyana and Venezuela. Rather, they are constructed in coastal savannas, where they are organized in groups or grids between the marshes and sand formations. Five principal organization schemas are defined for Guiana, each reflecting the level and degree of flooding as well as the local topography. Attempts to correlate these five schemas with the four types of raised fields have been unsuccessful. However, some trends do emerge, although they are not systematic.

Permanently flooded zones are frequently characterized by large constructions. These are elongated raised fields or large mounds arranged

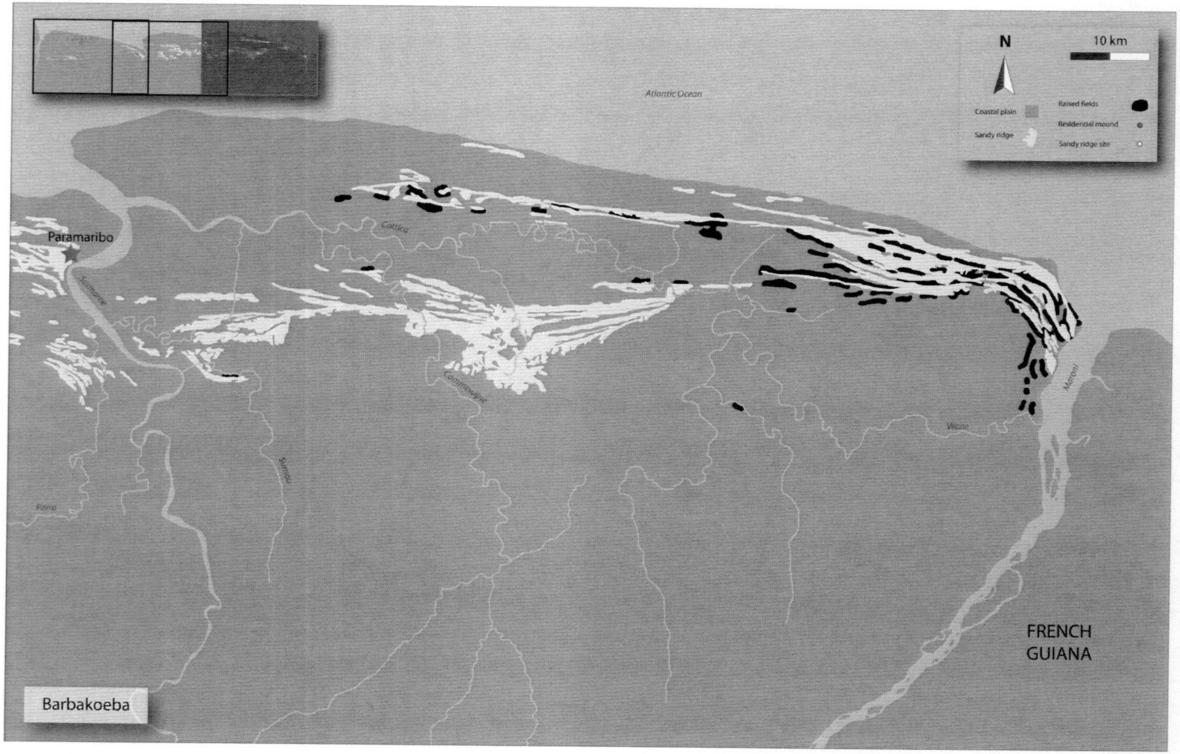

Figure 42. Raised-field complexes of Suriname, Barbakoeba sector (modified from unpublished Frans Bubberman draft, who added that his maps are not perfectly accurate).

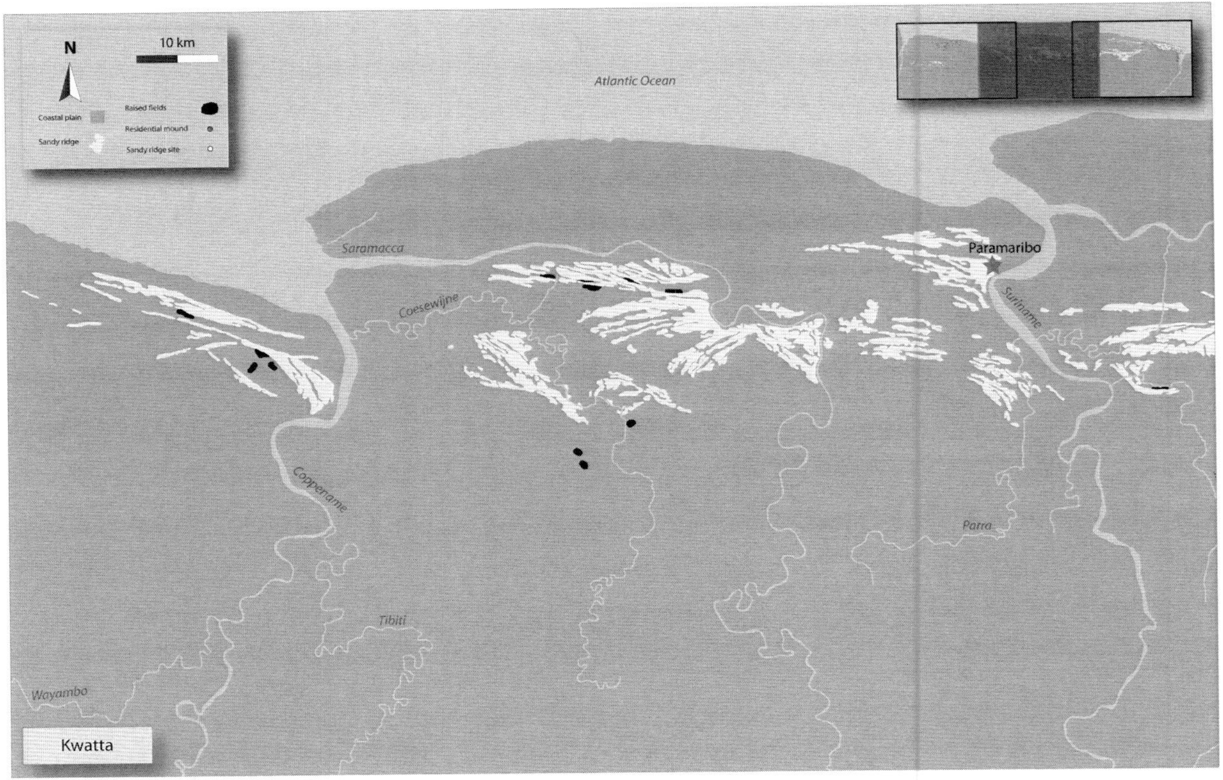

Figure 43. Raised-field complexes of Suriname, Kwatta sector (modified from unpublished Frans Bubberman draft).

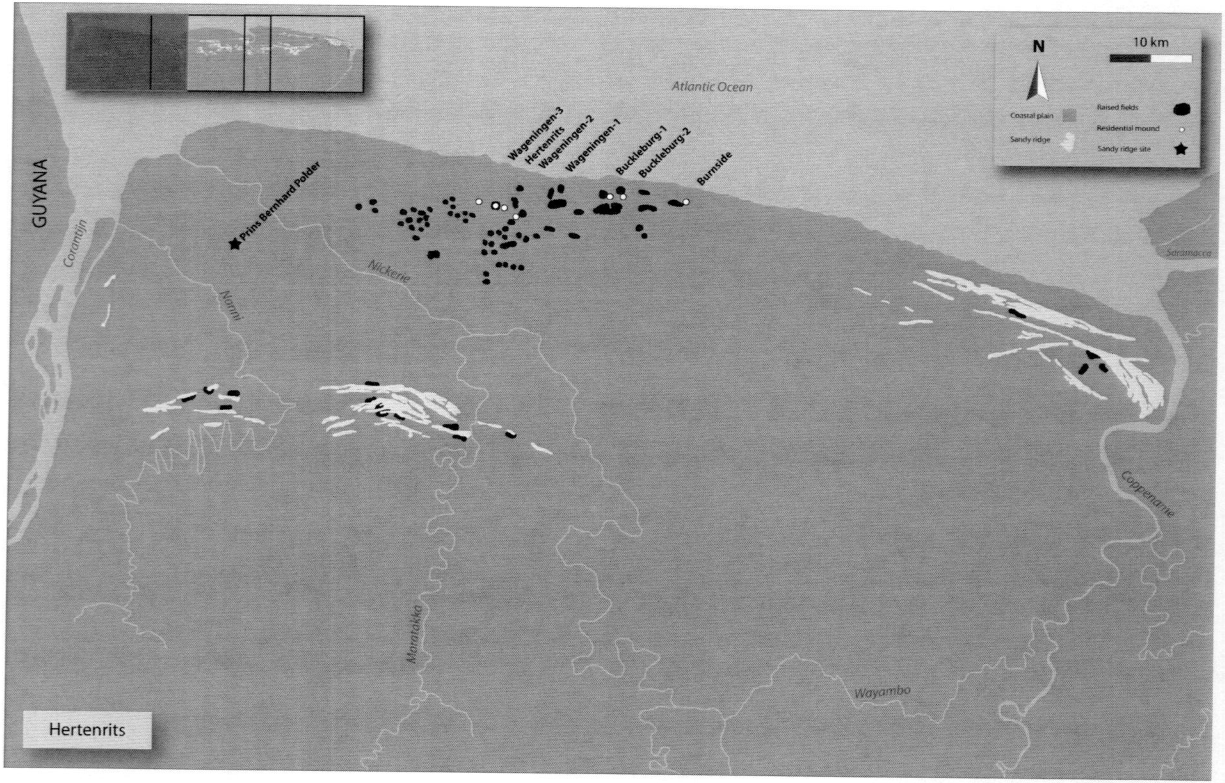

Figure 44. Raised-field complexes of Suriname, Hertenrits sector (modified from unpublished Frans Bubberman draft).

in a grid pattern. Such complexes have been recorded in permanently flooded depression in the Macouria and Awala sectors of French Guiana, as well as in eastern Suriname.

Seasonally flooded banks along rivers were cultivated in western Guyana. Parallel groups of elongated raised fields are arranged perpendicularly or in chevron patterns along rivers. They appear to function more often for irrigation than drainage. These structures are the most common type in South America (the most spectacular examples are along the San Jorge River in Colombia; see Plazas et al. [1993]), but can also be found in Guyana (although only in the Berbice and Canje Basins).

Thalwegs are the preferred locations for constructing raised fields because they provide highly concentrated fertile soils. They have been almost completely modified throughout Guiana. Medium mounds are the type most commonly found on thalwegs. These mounds follow the natural line of the thalweg and are distributed in sinuous formations across the landscape. They are abundant in French Guiana, and are the preferred locations for building raised fields around artificial habitation mounds in western Suriname.

The **edges of sandbars** are another important location for raised field construction because the slope allows for adaptations to changing flood levels over the course of the year. These zones are marked by the most complex examples of pre-Columbian agricultural technology; each mound is individually adapted to local hydrologic conditions. Thus, the shape and size of raised fields vary according to their location. This type of organization is common in French Guiana, particularly in the Kourou sector, but also in eastern Suriname.

The **seasonally flooded savannas**, completely dry in August, are the largest spaces cultivated by indigenous peoples. Almost the entire surface is modified; only the highest areas and the sandiest zones are left untouched. It is also notable that the most seasonally flooded areas (for example, the lowest-lying areas of the Grand Macoua savanna) are not used (Doyle McKey, personal communication, 2010). Across these vast, flat surfaces, raised fields are very similar and located closely together. Large, widely spaced mounds are reported only in eastern Guyana. But doubts regarding whether they are of anthropogenic origin remain unresolved. Elsewhere, the mounds are generally medium in size, although large and small mounds also may occur.

Thus, the morphology and the distribution of raised fields are linked to topography and water levels, with variability possible within a single complex. Furthermore, the mounds sometimes form homogeneous and independent groups extending a half-hectare each. Along the coast of Guiana, specific types of mound complexes dominate certain sectors. From west to east, seven principal modes of organization can be distinguished. Cultural or geographic reasons for these differences are not always apparent. Nevertheless, it seems that each sector corresponds to a specific archaeological culture rather than a specific environment.

The **Berbice** sector (Figure 36). Along the eastern part of the coast of Guyana the dominant pattern is a series of elongated raised fields oriented perpendicular to rivers. Also common are very large mounds, up to 2 meters high, widely spaced across the seasonally flooded savannas. Perhaps pertaining to the Barrancoid period, they would represent a different chronological phase.

The **Hertenrits** sector (Figure 45). In the recent coastal plain of western Suriname, complexes of elongated raised fields of varying sizes follow the contour of the thalwegs. The raised fields are either arranged in small concentrations or distributed in a disordered manner.

The **Barbakoeba** sector (Figure 46). In the east of the country, from the Suriname River to the Maroni, raised fields (large or medium) are principally distributed around Quaternary sand formations.

The **Awala** sector (Figure 47). Located on both banks of the Maroni and also on the Lower Mana, these consist of flooded depressions that have been modified by humans by the construction of large mounds or elongated raised fields arranged in grids.

The **Iracoubo** sector (Figure 48). Between the Organabo and Sinnamary Rivers of French Guiana, these are principally seasonally flooded savannas covered by large and medium mounds. There are also raised fields on the southern thalwegs, south of the Route Nationale No. 1 (National Road that runs east-west along the coast).

The **Kourou** sector (Figure 49). From the Sinnamary to the Macouria, elongated raised fields and mounds tend to be fairly randomly distributed along the contours of sandbars. South of the coastal plain, below Route Nationale No. 1, thalwegs are the preferred location and are generally associated with large and medium mounds.

The **Macouria** sector (Figure 50). From the Macouria River to the Île de Cayenne, although there is some preference for thalwegs, the savanna and flooded depressions are also used. The mounds are large and medium in size, and tend to be square.

The coastal stretch between the Hertenrits and Barbakoeba sectors is named Kwatta, where few raised fields have been detected. It is assumed that indigenous peoples cultivated the region's sandy shell ridges, which are supposedly more fertile than the coastal savannas.

Structures are generally arranged geometrically in a well-defined grid (although less symmetrical examples are occasionally noted). Our knowledge of the distribution of pre-Columbian earthworks on the coast of Guiana is still incomplete, but indicates that a great deal of the seasonally flooded zones to the west of Cayenne were transformed by pre-Columbian peoples.

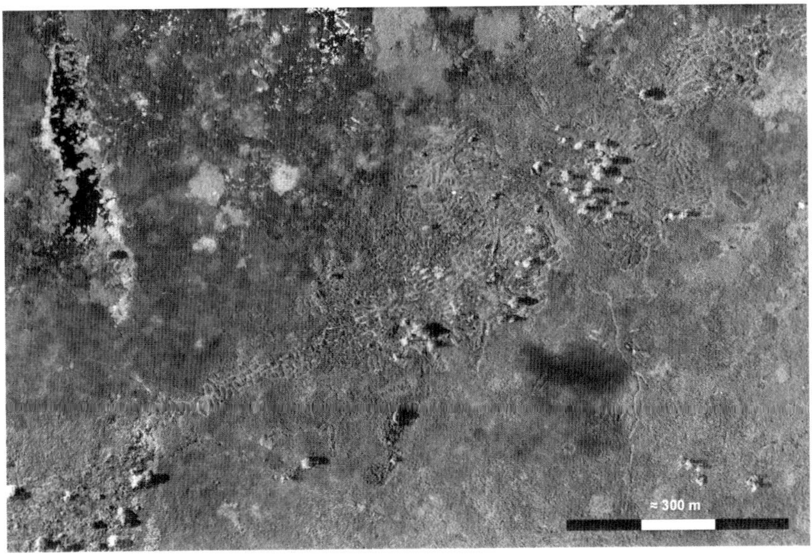

Figure 45. Raised fields of the Hertenrits sector (photo by Dutch Geographical Institute).

Figure 46. Raised fields of the Barbakoeba sector (photo by Stéphen Rostain).

Function

The elevation of the fields permits successful farming of land that has low agricultural fertility and is subject to periodic, prolonged flooding. The

Figure 47. Raised fields of the Awala sector (photo by Stéphen Rostain).

Figure 48. Raised fields of the Iracoubo sector (photo by Stéphen Rostain).

ancient coastal plain of Guiana is largely composed of marshes, which are unsuitable for cultivation without modification. The solution adopted

Figure 49. Raised fields of the Kourou sector (photo by Stéphen Rostain).

by indigenous groups was to construct mounds that reached higher than the highest flood level. This technique combines two principal functions: drainage of the soil and the concentration of fertile material.

The removal of soil from the matrix for constructing the mounds also produced drainage canals between the raised fields. The canals, in addition to their primary function of drainage, permit the elimination of harmful gases that limit root growth (Bernard Barthès, personal communication, 1989). This practice considerably reduces the risk of rot for the crops, particularly tubers.

Figure 50. Raised fields of the Macouria sector (photo by Stéphen Rostain).

The construction of raised fields in Guiana corresponded to a period of extremely humid climatic conditions. Between 1,300 and 800 years BP, intense rains provoked flooding in the Amazonian lowlands (Colinvaux 1989; Colinvaux et al. 1985). Pollen analysis conducted on the artificial habitation mounds of western Suriname demonstrate that they were built during periods of freshwater flooding (Versteeg 1983, 1985). The elevation of the surface of habitation mounds well above the present-day high flood stage for the region and that of Marajó Island may also be evidence that the flood level was higher during this past period. Drained land would therefore have been less available in the past. These conditions likely led to the development of raised-field techniques, at least in Suriname.

The second function of elevating mounds is to improve the structure of the soil. The clay soils of the marshes on the coastal plain are not well suited for farming, but the construction of mounds concentrates the highest-quality earth for planting. This technique also aerates the soil, which improves plant growth due to deeper penetration of roots and tubers. Restoring eroded mounds with sediment from between the fields rejuvenates the cultivated soil. Nevertheless, the application of fertilizers is important for increasing productivity. In the eighteenth century, Father Juan Gumilla ([1745] 1963) observed indigenous peoples constructing raised fields along the Middle Orinoco. He noted that they included dried vegetal material in the mounds. This same method of fertilization is still practiced by Brazilian and Haitian farmers who cultivate on raised fields. Findings by archaeobotanists José Iriarte and Jennifer Watling suggest pre-Columbian farmers also practiced this technique. Recovery of phytoliths characteristic of maize rachis, stems, and leaves in raised fields in the Grand Macoua savanna indicates that crop remains were incorporated in the mounds as fertilizer (McKey et al. 2010).

On the other hand, it is imprudent to include plant debris in mounds that are flooded year-round because organic decomposition in a waterlogged context could produce gleys that are toxic to plant roots (Groupe de Recherche et d'Échanges Technologiques [GRET] 1990). Gleys are soils that are deprived of oxygen due to waterlogging, which provokes anaerobic conditions and the reduction of iron. These conditions are asphyxiating and damaging for agriculture.

The physical and chemical potential of soils is essential to the fertility of raised fields. Pedological sampling of several complexes was undertaken in the 1980s to study the chemical properties of the soils. A cultural profile (observations of the structure and porosity of the soil used by plant roots, as well as the root structure of the vegetation) was conducted by Bernard Barthès (Rostain 1994a).

The soil of site K-IV, west of Kourou, has a high organic-matter content, particularly at the foot of the mounds, because it is submerged in water and the decomposition of organic matter is very slow. The levels of bases (e.g., calcium, magnesium, potassium, and sodium) and of phosphorus are more elevated at the foot of the mound than on top because these nutrients are most abundant in organic matter. The lower percentage of nutrients recorded at the top of the mound may be the result of leaching. Alternatively, nutrients may have been removed by the cultivated plants.

The soil of site M-II, in the Matiti savanna to the east of the Kourou River, has low organic matter content, although still higher (3 percent) than that of the soils on the upper portions of the coastal sandbars. These levels are due to the accumulation of organic matter in the matrix. The reserve of nutrients in the mounds is weak, but they are present. Even a small addition of organic fertilizers increases the soil's productivity, although the effect is short-lived. Archaeobotanical analyses made during our project in other raised-field complexes suggest fertilizers were indeed used in the past.

These observations need to be confirmed with quantified data. Nevertheless, it appears that these soils are some of the richest in the lowlands. Water control systems provide a further agricultural advantage. Today, the Wayampi of the Upper Oyapock, French Guiana, evaluate the texture of soil in terms of friability, but also its propensity for retaining water (which if excessive, can cause rotting of the plant's roots). Following these criteria, they distinguish several classes of agricultural soil (Grenand and Grenand in Turenne 2002).

The soils most favorable for agriculture are sandy; very sandy and dry; mixtures of sand with very permeable humus (considered ideal, particularly for manioc); black soils (composed of pure humus); coarse-grained sands; gravels containing small, round, and black concretions of manganese oxide; and coarse-grained gravels with more gravel than sand. These two latter types are the most difficult to cultivate.

The soils of medium utility for agriculture are red clays with medium to low humidity, red sands, bright red soils with abundant hematite concretions, impermeable clay soils, and sticky sand mixed with humid clay.

The low utility soils, used only as a last resort and always in conjunction with those of higher fertility, are damp soils with abundant earthworm castings and the sticky, clay soils that are even more damp and compact.

Soils that are unsuited for cultivation include the saturated mud found in low areas, very damp soils saturated with undecomposed dead leaves, and waterlogged silts full of decomposing leaves.

The farmers of raised fields surely must have developed similar distinctions among soils. This would explain why certain portions of the seasonally flooded savannas were not modified while others were densely covered with structures. The thalwegs, full of alluvial sediments, and the slopes of sandbars were apparently highly preferred over other locations.

Pedologists and archaeobotanists conducted sampling and measurements on the Grand Macoua and Organabo savannas (McKey et al. 2010). A comparison of phytolith assemblages and variations in the levels of stable carbon isotopes provides insight into the history of the landscape. Samples were collected from a stratigraphic column to a depth of 70 centimeters below surface. The phytolith spectrum at different depths provides an image of the dominant vegetal community at different points through time, with the deeper deposits older than the deposits closer to the surface. The analysis reveals two trends that evolved in a parallel manner. The first trend shows the increase of the relative proportion of grasses of the subfamily Panicoideae from the lower clay level to the upper, more organic level, and the decrease of the proportion of sedges (Cyperaceae) and herbaceous dicotyledons moving from the lowest level to the surface. The second trend indicates fairly homogeneous vegetation in the lower levels transitioning to more differentiated vegetation between the mounds (grasses) and the matrix close to the surface (Cyperaceae). These results are confirmed by isotopic analysis of the organic material in the soil, which shows a transition from a homogeneous isotopic signature in the lower levels (indicating a higher proportion of C_3 plants) to a heterogeneous signature near the surface, with the mounds displaying a strong signal of the contribution of C_4 plants (the greatest contributor being the grasses of the Panicoideae subfamily) and the matrix dominated by C_3 plants (i.e., herbaceous plants such as *Heliconia*, Marantaceae [arrowroot family]) and a few grass species, along with most sedges (Cyperaceae). The modern vegetation at the site demonstrates the same pattern, with a stronger proportion of C_4 plants on the mounds and C_3 plants in the matrix. The change in the vegetation communities can be interpreted as the result of a change from a naturally flat and flooded savanna to a savanna scattered with mounds, where humans introduced heterogeneity (see also Renard et al. 2011).

The principal physical constraint is access to water and control of its circulation. Indigenous people deliberately located their mound complexes between low-lying areas and sandbars. Nevertheless, this precaution did not always protect the mounds from flooding, and it was sometimes necessary to surround certain groups of fields with drainage canals. The distribution of elongated raised fields in grids also reflects a concern for water control. In several complexes, the fields follow the downhill slope of the sandbars, which was likely to facilitate drainage. Conversely, the elongated raised fields located upstream (in a less humid environment) are distributed perpendicular to the slope to favor water retention (Figure 39).

The mounds serve a double function. First, they protect plants from flooding because they elevate the planting surface above the flood level and facilitate drainage because the soil cannot absorb all of the rainfall during the wet season. Second, they concentrate organic material (residues of harvests and weeds) for the more demanding crops, such as maize or yams. The mounds also permit aeration of the soil. Physical and chemical constraints could be easily overcome by minimal fertilization and efficient water control. In any case, locating raised fields on the edge of flooded, seasonally flooded, and dry zones reflects a conscious decision-making process by indigenous peoples. Finally, compared with highland slash-and-burn cultivation, it seems that raised fields allow for better regeneration of soils and are more sustainable over the long term.

Detailed Study of Raised-field Complexes

Several mound complexes have been carefully studied using stereoscopy and precise topographic mapping, as well as geomorphologic, pedologic, botanical, and ecological sampling. The K-VIII complex was selected for high-precision topographic mapping during the dry season of 1989. More than a thousand planimetric points were recorded for an area that measured 650 by 500 meters. Most of the elongated raised fields and large and medium mounds rise approximately 50 to 60 centimeters above the surface of the surrounding marsh. The careful organization of the mounds is noteworthy (Figure 51). The complex forms a crude "S" shape, following the natural contour of the sandbar. The raised fields are located on the slope between the year-round flooded zone and the sand formation's dry summit. The largest rounded mounds were constructed in the center of the marsh, the most flooded area. Upstream from these mounds are parallel rows of elongated raised fields that follow the natural slope to facilitate drainage. Behind these mounds, near the summit of the drier sandbar, the groups of parallel, elongated raised fields are arranged perpendicular to the slope to retain water, which is less abundant in this location. This is a prime example of indigenous agricultural engineering.

During the "Archaeology and ecology of the coastal savannas of Guiana" project, pedologists demonstrated that the sediments used to build the mounds had been carefully selected. The paleosol A horizon that is rich in humus and buried in the raised fields is at the same level as the humus that occurs outside the mounds. Above this buried A horizon is a fine layer of subsoil clay 5 centimeters thick. This stratigraphic sequence demonstrates that the bases of the mounds were constructed on the upper surface of the surrounding marsh. This level is covered by the mound's fill, a dark sediment that is rich in organic material and close to 60 centimeters thick (Figure 51). This sediment does not come directly from the marsh, but was transported from elsewhere. Although a humus-rich horizon is present in the center of the sandbar, this horizon is not present in a large portion of the circumference of the summit. The sterile subsoil appears on the cleared surface, where vegetation is rare compared with the surroundings. This contrast cannot be attributed to a natural phenomenon. It is likely that indigenous people scraped the richest soils from the sand formations to build the agricultural mounds.

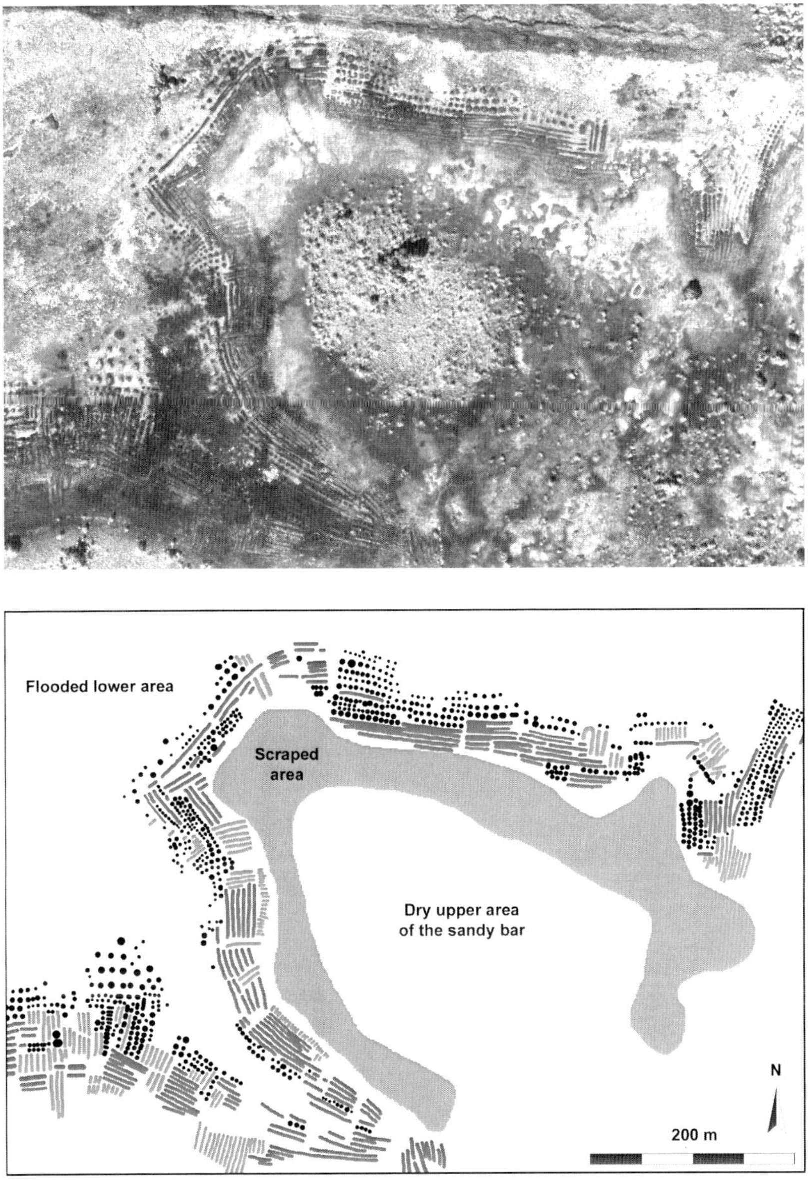

Figure 51. Stereoscopic interpretation of K-VIII complex, west of Kourou, French Guiana. The gray area corresponds to the soil probably scraped by Amerindians to built the raised fields (photo, IGN, 1987).

They appear to have used shovels and baskets to transport the earth over a distance of less than 100 meters (McKey et al. 2010). Pedological analyses confirm the similarity between the samples collected from the humus-rich level of the sandbar and in the nearby raised fields.

The pair of photographs taken in 1955 from the Intstitut Géographique National (IGN) GUY-003/100 (Nos. 369 and 370) presents a complex from the Matiti savanna, east of Kourou (Figure 52). The remains have been obliterated by alluviation. As elsewhere, the sand level is approximately 4 meters high. The summits of granitic hills partially buried by

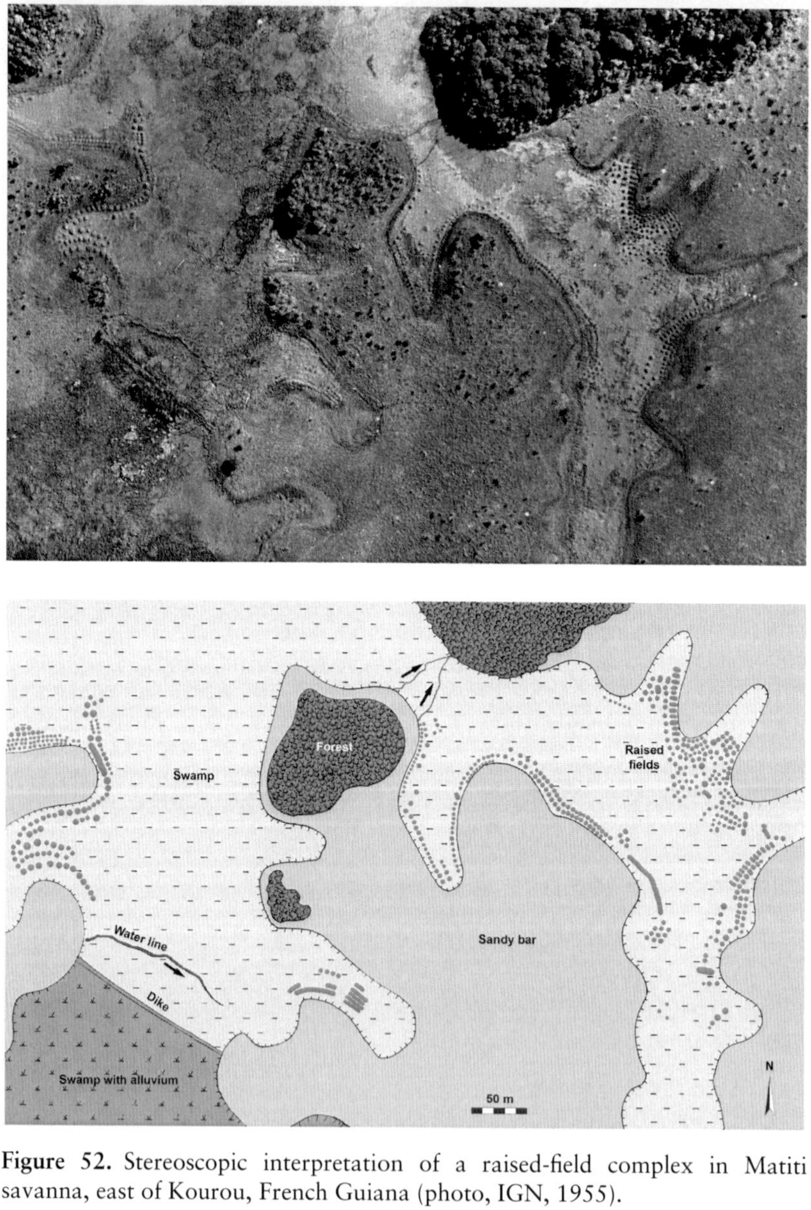

Figure 52. Stereoscopic interpretation of a raised-field complex in Matiti savanna, east of Kourou, French Guiana (photo, IGN, 1955).

the alluvial sediments emerge from the sand formations. One notes the absence of colluvial deposits around this relief, which does not appear to have greatly influenced the configuration of the lowlands.

Mounds flank the perimeter of the sandbars, with a width of 20 to 50 meters. It is likely that the entire slope was previously modified, but that many of the structures have been buried by alluvial action. Transversal lines, without relief, are found on the slope and could be interpreted as the traces of water runoff. However, their origin remains enigmatic because they originate from prominences, rather than hollows, in the contour of the sandbars. It is possible that humans dug these alignments. This is clearly the case for the earthen embankment that appears to have been designed by humans to protect the seasonally flooded area to the south (Figure 52). If this hypothesis is correct, the construction of the embankment would have allowed pre-Columbian farmers to control the flow of water in this space to increase the efficiency of the mounds. This embankment appears to have provoked the development of a very significant alluvial deposit upstream, particularly on the expanse of the southern marsh, which has almost completely obliterated all of the ancient modifications. This process likely took place over several centuries.

We searched for these structures in during the 2008 survey, but although the raised fields were easy to locate, the levee and the runoff lines had disappeared, probably due to recent alluvial and colluvial action.

In K-IV complex, located in the northeastern part of the Centre Spatial Guyanais, the mounds extend over approximately 500 square meters and follow the contour of the sandbar. The poorly drained soil is completely submerged during the rainy season from the flooded marsh to the edge of the rainforest. The site is made up of 128 large, circular mounds (from 3 to 4 meters in diameter) and three elongated raised fields (two of which measure 30 meters long and 3 meters wide, while the third is 9 meters long and 3.5 meters wide). Their current height varies from 30 to 70 centimeters. Two tests were carried out with a bucket auger by Bernard Barthès to study the stratigraphy of the mounds, one at the foot of a mound and the other at its summit (30 centimeters high).

In the matrix, from the surface down to 35 centimeters below surface, a layer of black silt that exudes a sulfurous odor corresponds to the Demarara Formation. Between 35 and 45 centimeters below surface appears a white sandy level with dark yellow parts, related to the Coswine Formation. Beginning at 45 centimeters below surface, the white and dark red marbled soil becomes increasingly more clay-rich and drier, finishing as pure Coropine clay.

The mound is made up of Demarara Formation sediment, as demonstrated by the test excavation at its summit. The soil is more aerated and drier; thus, the sulfurous odor (generally associated with poorly drained soils) is not present. At the top of the mound, the proportion of sand is slightly elevated, probably due to the downward leaching of fine particles (clays and silts).

All of the samples are strongly acidic, which is normal in a hydromorphic environment. Stratigraphic study and granulometric analyses demonstrate that the mounds were constructed using sediment derived from the surrounding matrix. This finding is similar to those for sites in western Suriname, where raised fields were constructed directly over the Moleson clay deposits or over fine, peaty levels (Versteeg 1985).

In the K-IV complex west of Kourou (T-I, north of Montagne des Pères, and M-III at Malmaison) a clear difference in the vegetation is apparent between the humid matrix and the tops of the mounds. The plants recovered by botanists George Cremers and Michel Hoff were primarily those that are adapted to a wide range of conditions, which are common in the savanna (Cremers 1990), and several ruderal species of anthropogenic origin. The marshy soil is populated by Cyperaceae (three species recovered), whereas the mounds are covered with secondary vegetation, principally grasses. The presence of anthropogenic species on the mounds studied would have resulted from human activity, estimated to span several dozen years. As these mounds were close to habitations, and were relatively easy to access, their vegetation appears to have resulted from a Guianese custom. Each year, at the end of the dry season, fires are set at several points across the savanna. The dry vegetation allows the fires to spread rapidly and cover vast distances. As one observes the fires advance, it is clear that the thick scrub that covers the mounds burns quickly, while the vegetation of the matrix that remains humid year-round is only partially destroyed by the fire. Furthermore, the vegetation of the matrix recovers much more rapidly than that of the mounds. These repeated burns explain the presence of young vegetation on the mounds.

Certain groups of raised fields are currently covered by the forest. These mounds are of medium size and irregularly shaped, distorted by tree roots. If the forest was already present in these zones during the construction of the mounds, clearance would have been necessary. However, this hypothesis remains unlikely, and it seems more probable that these structures were originally constructed in an open marshy zone that was later infilled by erosion and then colonized by the forest when fires were less frequent.

Dating

Dating raised fields directly is challenging because they are constructed from material transported from surrounding areas and thus contain little datable material. The anthropogenic remains that are sometimes recovered from the mounds are generally not useful because their presence does not necessarily correspond to the period of usage. In addition, they may also consist of material that predates the mound but was brought there during its construction, or conversely, material abandoned on the mound more recently. Although these possibilities may be unlikely because the seasonally flooded savannas were little exploited prior to the pre-Columbian earthworks and generally ignored by European colonists, they cannot be dismissed. Thus, the few nondecorated sherds (apparently of Barbakoeba style) discovered during the excavation of a raised field at the site of Wane Creek in Suriname could correspond well to these structures' architects (Arie Boomert, personal communication, 1993).

There are several indices for deductive dating. The recent coastal plain of western Suriname was almost never used by Amerindians, Europeans, or Creoles from the Conquest until the 1950s. Therefore, the remains and ecofacts found there are most likely of pre-Columbian origin. Furthermore, the raised fields of this region have always been organized around artificial residential mounds that are clearly indigenous and associated with the Barrancoid or Arauquinoid periods.

The discovery of a wooden shovel in the peat of the Arauquinoid site Prins Bernhard Polder, dated to the thirteenth century AD, supports this chronological association. This tool, perfectly adapted for excavating clay to construct mounds, was no doubt abandoned during the period of construction, and thus represents a useful chronological marker. Elsewhere in the tropical lowlands of South America, mound complexes are also associated with constructions that are clearly pre-Columbian and located in areas recently taken over by colonists. Nevertheless, those circumstances represent only indirect evidence of the antiquity of the raised fields and are insufficient for secure dating.

Organic material extracted from the tops of the paleosols buried under the mound complexes of Bois Diable and K-VIII, west of Kourou, has been dated using ^{14}C. The ages obtained are, respectively, 760 ± 40 years BP (calibrated, 670–700 years BP) and 1,010 ± 40 years BP (calibrated, 920–950 years BP) (McKey et al. 2010). Other ages are 1,060 ± 30 years BP (calibrated, AD 898–1022) in Bois Diable and 620 ± 30 years BP (calibrated, AD 1289–1404) in the Matiti savanna. These dates correspond to those obtained from the sites of Barbakoeba on the west coast of French Guiana and from the eastern coast of Suriname. It is therefore reasonable to attribute the construction of the raised fields to the Barbakoeba groups that inhabited the sand ridges bordering the mound complexes.

In western Suriname, the mounds are clearly associated with Barrancoid or Arauquinoid mounds of the Early or Late Hertenrits culture.

A Landscape Sculpted by Humans

Complementary Earthworks

Although raised fields constitute the principal landscape modification used by indigenous peoples, other transformations to the environment are known. Canals were frequently associated with agricultural constructions to improve drainage. However, not all construction served an agricultural function. Streams were diverted and canals were excavated to facilitate their circulation. Roads were elevated to cross flooded areas. Finally, enormous clay mounds were constructed as foundations for villages.

Canals

Stereoscopic interpretation reveals a tremendous number of excavated features, both natural and anthropogenic. These consist of irregular lines created by water flow or rectilinear canals. It is sometimes difficult to determine which are anthropogenic in origin because these earthworks can be easily confused with features caused by natural runoff. Due to their horizontality, the seasonally flooded savannas are characterized by a dense and disordered hydrologic pattern. However, these natural networks are too random to aid with drainage of the raised fields. Thus, it was necessary for indigenous peoples to excavate ditches to redirect excess water, which they often accomplished by simply modifying segments of the existing natural channels. In other cases, they excavated straight and wide drainage canals. This latter technique was also practiced by European colonists, who constructed long canals for the polders, but also as routes for communication. Leroy Canal, which stretches for almost 11 kilometers to the west of Kourou, is an example of these colonial constructions. It is therefore often difficult to identify with certainty the nature and origin of the drainage routes observed. Four types of hydraulic modifications are distinguished.

Enclosed ditches are linear features that follow an irregular course, perpendicular to the natural flow of water (Figure 53). They often connect two areas of high ground and enclose a group of raised fields. These small, artificial features measure from 1 to 2 meters wide and several dozen meters long. They provide protection from flooding during the rainy season. They are relatively narrow and shallow at their extremities, but often form a basin at the center point. These ditches therefore not only divert excess water in the rainy season, but they also provide a reservoir of water in the dry season.

Canals are rectilinear ditches, wider and deeper than enclosed ditches. They measure between 50 and 600 meters long and between 2 and 5 meters wide. They often enclose a seasonally flooded space and appear to control its drainage. Beyond their drainage function, the ditches and

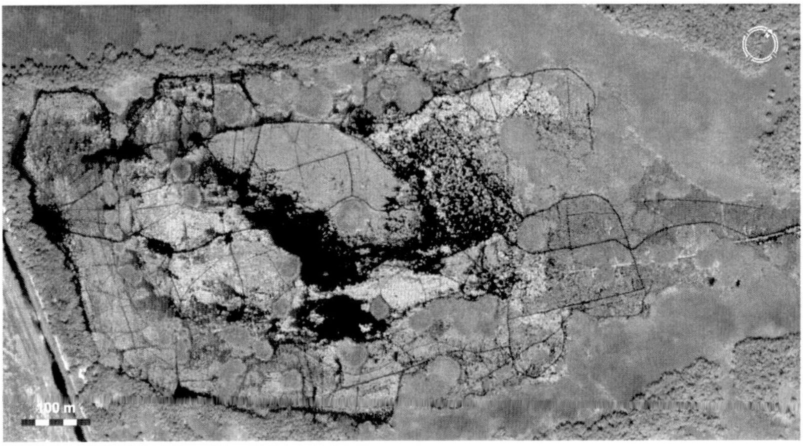

Figure 53. Net of canals near the Wageningen-3 mound, Suriname (modified from Google Earth). Image © 2012 DigitalGlobe.

canals can also serve as reservoirs of water during the dry season, and as permanent fishponds. Nevertheless, this type of canal is relatively rare. In French Guiana, in the Kourou sector, they are more frequent to the south of Route Nationale No. 1 than to its north. This distribution is probably due to greater problems with drainage in the southern zone, which is elevated slightly higher than the northern plain. Canals are less common toward the marine shore, where they would be counterproductive because the clay soils become impermeable once they are saturated. Also, the construction of a dense drainage network could provoke a rise of brackish water, which would irreversibly render the soil uncultivatable (GRET 1990).

Raised fields are constructed and organized based on consideration of the seasonal levels and circulation of water. Ditches are constructed to control water flow.

Reservoirs are excavated areas that function to conserve water and/or for pisciculture. They are not formally recorded in French Guiana, but occur to the east of Suriname (Boomert 1976).

Dug paths and the channeling of streams are similar to canals, but they are much smaller in their dimensions and their usage is radically different. Rather than being used for agriculture, they facilitate circulation. The most notable example is associated with the artificial settlement mounds of Hertenrits, where a network of these linear furrows forms a grid 10 kilometers in diameter (Figure 54). Indigenous peoples have drastically changed the course of small natural streams that cut across the plain. Most of the time this involves altering the course of water flow to render it more rectilinear. In other cases, two existing

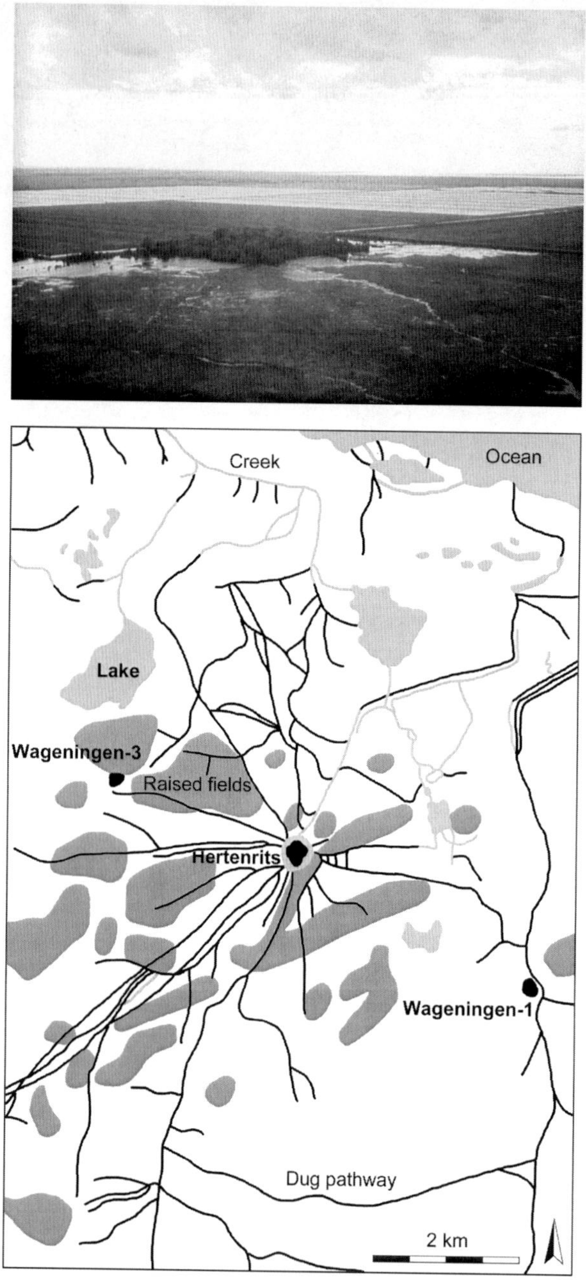

Figure 54. The Arauquinoid mound of Hertenrits, Suriname (photo by Dirk Geijskes), and stereoscopic interpretation of Hertenrits area (drawing by Stéphen Rostain modified from Boomert 1980b:73).

creeks are linked by a passage. Completely human-made channels occur as well. The Hertenrits mound is surrounded by a large spider-web pattern of dug paths. Some lengthen the course of natural streams, while others run parallel to creeks. Many extend out from residential mounds and link with raised-field complexes or secondary domestic mounds. These sunken paths serve a double function. During the dry season, they mark paths to arrive at a specific location. These ancient networks resemble those associated with modern indigenous villages today. During the rainy season, when the plain is flooded, the same paths can be used as canals for canoes. The flooding is not very deep across the plain and boats thus risk running aground. The sunken paths serve as canals that permit the navigator to direct his boat safely across the landscape.

In eastern Suriname, the flow of creeks has also been modified to aid the circulation of boats (Frans Bubberman, personal communication, 2005). Across Guiana, canal reaches are common on the bends of rivers. These are canals that pass through the forest to facilitate passage of canoes and shorten the overall route by not requiring the craft to follow the longer, natural course of the river. In some cases, the shallow canals were constructed to connect two points of the river.[1] Therefore, it is possible that some of the canals that connect higher parts of the coastal plain of Guiana served as routes for canoes.

In French Guiana, two ancient human-constructed canals cross the road No. 22 linking Mana to Awala (Cornette 1987). They pass through the *chenier* on which the road is constructed and enter the marshes on both sides of the sand formation. The first canal cuts the road at 8.6 kilometers east of the village. Its length is estimated at 2 kilometers, but it is not perfectly straight because there are two slight changes of its orientation from 10 to 30 degrees east. At its intersection with the road, it measures 21.4 meters wide at the top of its walls (a slope of 25 degrees to the northwest and 31 degrees to the southeast). It is 4 meters at its base and its maximum depth is 3.15 meters. It has a concave section and a relatively flat base. When it enters the forest to the south, the width of the canal diminishes to an average of 8 meters. The second canal, 12 meters wide, crosses the road perpendicularly at 5.2 kilometers from the first and is located several dozen meters east of raised fields. Another canal, 5 meters wide and approximately 1 meter deep, is noted close to the ancient Kali'na village of Coswine. According to the indigenous people of Awala, these canals were related to the site of *Nieku dupo* ("Where the poison root for fishing is found"[2]), which today no longer exists. The myth of *Sibalalimbo* ("the man dressed in iron") tells that the Kali'na of this village had dug sunken paths to quickly assemble their warriors during conflicts with the English during the Colonial period.

Causeways

The inverse of the dug paths are the raised paths, or causeways. Stereoscopic interpretation has identified artificial earthen levees in several highly flooded areas of French Guiana, close to the modern shore. Five constructions of this type have been recorded. These levees cross the marshes perpendicular to the shoreline. Three link together two *cheniers*, while the final departs from a *chenier* and ends in the mangroves along the shore. North of Trou Leonard, west of Kourou, one levee measures 430 meters long and 8 meters wide. North of the Matiti savanna, toward the area called Guatemala, another levee measures 330 meters long and 5 meters wide. The levees of the Corossony savanna and Pointe Charlotte were surveyed in 2006 and 2009, but it is difficult to determine if the earthwork at Corossony is of indigenous origin or more recent.

The Amerindian origin of the causeway at Pointe Charlotte can be more easily demonstrated (Figure 55). It is clearly visible on the IGN aerial photographs from 1987, but the southern part was almost totally destroyed 20 years later. Local hydrological conditions have likely changed, which would explain the disappearance of half of the structure. The causeway would have linked the *cheniers* of Colline Sable and Koroni across a flooded depression. During archaeological survey in 2008, Laure Déodat and Lydie Clerc attempted unsuccessfully to reach this causeway from Colline Sable (Rostain et al. 2008). Despite GPS tracking based on the IGN aerial photos, all attempts led into the marsh, although an alignment of shrubs observed in the distance to the north was suspected to mark the causeway. In 2009, a new attempt was undertaken using the northern *chenier* of Koroni as the departure point. This time the causeway was located, confirming that the row of shrubs that had been previously noted did indeed mark its location (Clerc and Déodat 2009). As additional proof, the sets of GPS points taken in 2008 and 2009 both indicate the same point of departure for the causeway. These findings confirm that the southern end of the causeway had been correctly located in 2008, but that this portion of the structure had been previously destroyed while the northern half remained preserved.

The earthen levee, running northeast to southwest, is characterized by a convex profile and measures 450 meters long and 3 meters wide. It presents several breaks 1.3 meters wide that cut across its total width, probably as a result of natural water infiltration. It does not appear that these breaks represent human-constructed canals designed to aid with the circulation of water or to avoid flooding of the causeway. The deepest portion of the marsh, toward the middle of the causeway, is used for fishing by the inhabitants of the region and appropriately called "Fish Hole." The levee thus facilitated access to a prime fishing location as

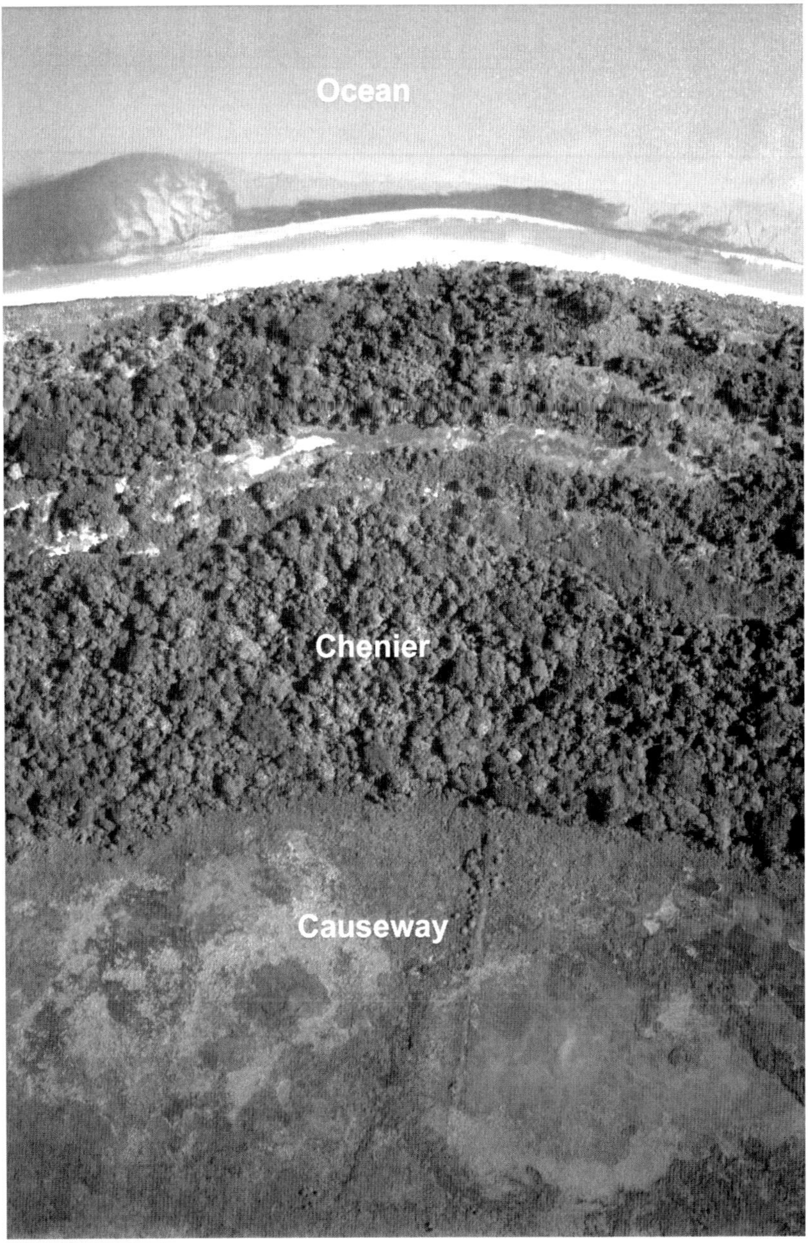

Figure 55. Causeway of Pointe Charlotte, west of Kourou, French Guiana (photo L'avion Jaune).

well as provided an effective blind for hunting birds. The marsh is too deep to cross on foot, and the causeway allowed direct access to the shore from the southern *chenier* without having to pass around the ends of the sandbar, substantially shortening the trip. This assumes that the Koroni *chenier* ran along the coast during Colline Sable's occupation, which is highly probable.

Although it is not completely certain that indigenous peoples constructed this earthwork, the available evidence suggests that this is the most likely explanation. There is no mention in the region's historical archives of such a construction being carried out by slaves, convicts, or Creoles. The former inhabitants of the Kourou savannas have no recollection of the levee having been constructed. Furthermore, the remains of a Creole settlement on the Colline Sable *chenier* are located on the northwest end of the sand formation, which provides access to the Anse and the beach without the need to cross the deep marsh. Thus, the causeway would not have served a vital function for the Creole occupants. Testing and excavations have shown that the Colline Sable *chenier* was completely occupied by indigenous people during the pre-Columbian period, a settlement that would have extended over almost 3 kilometers. For those living on the center of the *chenier*, the marsh would have been impossible to cross; a long detour to the end of the bar would have been necessary to reach the shore. However, the construction of a raised pathway several hundreds of meters long would resolve this problem and pose little difficulty for a large group accustomed to farming on artificial raised fields. Although no direct proof has yet been obtained regarding the architects of this structure, the clues point firmly toward indigenous peoples.

Habitation Mounds

There are almost no *cheniers* in the recent coastal plain of western Suriname and eastern Guyana. Thus, indigenous peoples had to adapt to construct settlements. Due to the lack of sufficient dry land for construction, they constructed immense, circular clay mounds above the flood level. Arauquinoid populations were accustomed to wetland environments and skilled in their modification. Several artificial mounds and other earthworks are found within the Arauquinoid sphere of influence in the Venezuelan *llanos* (Redmond and Spencer 2007). Versteeg's (1985) doctoral thesis at the University of Leiden in the Netherlands presented a study of mounds in western Suriname. In Suriname, of the 15 potentially anthropogenic mounds recorded by van der Heide (1973), seven are clearly ancient indigenous villages, two do not contain archaeological materials, two were circular mounds produced by the growth of *Avivennia* mangroves, and the remaining four were not visited (Versteeg 1985). The mounds in eastern Guyana are less well known due to a lack of archaeological work, but most of them appear to belong to the Abary culture, a Barrancoid tradition defined by Clifford Evans and Betty Meggers (1960).

The Barrancoid peoples were the first to construct settlement mounds in Guiana during the first few centuries of the Common Era. Analysis and dating of pollen from peat layers indicates that between AD 300 and 1000, freshwater conditions prevailed on the western plain of Suriname (Versteeg 1985). It was at this time that Barrancoid groups began mound construction. The mounds were constructed in stages, as is evidenced by the succession of light and dark layers in the stratigraphy. The former correspond to the foundational fill, while the latter represent the actual remains of the occupation.

Buckleburg-1 is a circular mound that measures 139 by 137 meters in diameter. Its mean altitude is 2.7 meters asl, or 1.5 to 1.9 meters above the level of the surrounding marsh. Humans arrived in the area around 1,835 years BP, and construction began at 1,735 years BP. The mound was abandoned by the Barrancoid people shortly after 1,315 ± 55 years BP.

Buckleburg-2 is slightly larger and taller than Buckleburg-1, located 2.1 kilometers to the west/southwest. The two sites share similar ceramics. Furthermore, the two mounds are built along the same river.

Wageningen-1 is a slightly oval mound that measures 130 by 100 meters in diameter and has an altitude of 2.25 meters asl, or close to 1 meter above the level of the marsh. Its construction began at 1,315 ± 20 years BP, and its occupants left after 1,130 ± 45 years BP. The ceramic style is defined as Early Hertenrits, which corresponds to the Early Arauquinoid period.

Wageningen-2 is a small oval mound located 1.6 kilometers southeast of Hertenrits and is associated with this culture. It measures 60 by 40 meters in diameter and is 40 to 50 centimeters high.

Wageningen-3 is situated 2.8 kilometers to the northwest of Hertenrits. It measures 80 by 60 meters in diameter and 40 centimeters high. Like the two previous mounds, it is located in the territorial network of the Hertenrits culture.

Burnside is the easternmost mound in the region. Oval in shape, it measures approximately 100 meters in diameter and rises 50 to 75 centimeters above the level of the marsh. The ceramics collected from the mound are associated with the Hertenrits culture.

Joanna, on the banks of the Canje River, is the only well-described mound in eastern Guyana (Goodland 1964). Surrounded by a ditch from which the construction material was derived, it measures 90 meters in diameter and 2.5 meters high.

Hertenrits, 6 kilometers south of the modern shoreline, is the largest of all the known residential mounds and has the longest occupation. It was named "sandy ridge of the deer" (*hert-en-rits* in Dutch) by the hunters who used it in the past. Oval in shape, it measures between 200 and 320 meters in diameter and averages 2.9 meters above the level of the marsh (Boomert 1980b). The ceramic sherds recovered in the different levels are divided into two successive styles, Early and Late Hertenrits (Versteeg 1985). The mound was inhabited between 1,265 and 935 ± 35 years BP, but the stylistic change at the end of the sequence is interpreted as the result of the arrival of a new influx of population from the Orinoco around 1,000 to 900 years BP.

The methods used by indigenous peoples to build earthen constructions long remained unknown. Fortunately, the extraordinary discovery of one shovel made in the marsh close to the site of Prins Bernhard Polder is very informative. The tool made from wood of green ebony (*Tabebuia serratifolia*, Bignoniaceae) is 72 centimeters long, with a flattened end that is sharp and curved, and a cylindrical handle (Figure 56). This shovel was dated to 790 ± 30 years BP, or 693 to 733 years BP calibrated (i.e., the Late Arauquinoid period) (Versteeg 1985). It was no doubt used at the time of the earthworks. In 1745, Father Juan Gumilla ([1745] 1963) reported that the raised fields of the Venezuelan *llanos* were built using wooden shovels, which he compared to the large wooden swords of the indigenous peoples.

Similar wooden spades still existed into modern times among certain Amazonian groups, such as the Ashluslay of Paraguay (Nordenskiöld 1929). This type of tool is also present in the pre-Columbian sites of Mexico, as is demonstrated in the collections at the Templo Mayor Museum in Mexico City. The tool that is most comparable is that found among the Floup of Senegal, who use wooden spades called *kayendos* (Figure 56) outfitted with a sharp metal end (Montoroi 1996; Pélissier 1966) to cut and extract blocks from clay. Archaeological excavations have shown that the Hertenrits mound was constructed by the stacking of rectangular clay bricks, probably through the use of spades similar to that recovered from Prins Bernhard Polder.

Two other fragments of tools were found at Prins Bernhard Polder. These are tortoise shell blades. The more complete, 10 centimeters long, contains two perforations and one central groove on each side (Versteeg 1985, 2003). It is possible that these are remains from shovels or hoes that were also used to build the earthworks. Unfortunately, the preservation of wood and bone in tropical environments is far from ideal, and it is unlikely that many more tools used in the construction and maintenance of raised fields will be recovered.

Well-defined Boundaries

Stereoscopic interpretation, archaeological excavation, and the analysis of the natural and anthropogenic environments of the coast of Guiana have revealed significant impact by pre-Columbian populations on their landscape. These transformations were accompanied by rigorous management of the territory and the organization of many small parcels.

Figure 56. Top photo and drawing: Arauquinoid wooden shovel from the Prins Bernhard Polder site, Suriname (photo by Stéphen Rostain). Bottom photo: the modern *kayendo* from Senegal, very similar to the pre-Columbian tool (photo by Jean-Pierre Montoroi).

Furthermore, the sociopolitical scope of different settlements seems to have varied, with some exercising more power than others.

One of the most significant criteria to determine the existence of chiefdoms is the hierarchical organization of sites based on their political importance. This aspect has been, to date, little studied in Amazonian archaeology. I have thus attempted to evaluate whether Arauquinoid sites fall into a hierarchical model.

In the floodplain of the Guianese coast, the *cheniers* offer ideal locations for villages, which are arranged along the length of these sand formations, extend parallel to the coast, and are connected to one another by paths. Although each of these settlements was limited by the narrow width of the sandbars (less than 200 meters), they spread over several hundreds of meters, at times even over many kilometers.

The Barbakoeba site of Sable Blanc, on the western part of the coast of French Guiana, stands out because it combines three different functions: residential, funerary, and agricultural, the agricultural site located some hundreds of meters south of the settlement. The residential zone is located at the top of the sandbar, with the houses constructed at the highest and driest point. The settlement was arranged in a linear form, punctuated by open spaces, along the length of the *chenier*. The site is found on the left bank of the Iracoubo River, just south of its mouth. This location corresponds to a recurring model evident for the ancient coastal Arauquinoid villages. The largest sites were systematically located a short distance to the west of a large river and close to the coast. This is the case for the sites of Bois Diable and Crique Jacques in French Guiana (Rostain 1994a), and Kwatta-Tingiholo and Peruvia-2 in Suriname (Versteeg 1985). Access to the ocean was important because the farmers of raised fields also depended heavily on marine resources. The southern slope of the *chenier* was reserved for funerary activities, and the absence of domestic structures supports this interpretation. Two distinct groups of urns with similar dimensions appear in this sector, and it is likely that other such groups can be found in the rainforest farther to the west. These concentrations of urns, simple but with different associations of burials and offerings, could correspond to clans or families within the village. In this case, the cemetery would have been organized very precisely. The drainage canals that cross the necropolis indicate that it was necessary to manage an excess of water at certain times of the year. An artificial embankment 4 meters wide and 1.5 meters high surrounds the cemetery. It could have served to protect the necropolis from flooding when the level of the southern marsh rose during the rainy season. In the lowest portions to the south and west, less than a kilometer from the site, there are raised-field complexes that cover the savannas. As in the other sites of Quaternary sandbars, the preferred placement of raised fields is in the marshes south of the settlement.

Thus, Arauquinoid sites of the *cheniers* demonstrate a specific organization of the settlement and its surroundings. The boundaries are well defined, and the communities make the most advantage possible of their space. It is difficult to believe that these modifications to the territory resulted from small, independent groups. In addition, notable nonagricultural earthworks were sometimes created. These include embankments, causeways that cross the marshes, and canals or reservoirs (possible fisheries).

These findings are even more convincing for the artificial mound sites of the western part of the coast of Suriname. Pollen samples collected around these constructions show that they were located at the interface of freshwater and saltwater environments (Versteeg 1985). The areas around the mounds were flooded with freshwater at the end of the dry season. This permitted the inhabitants to exploit resources from marine, brackish, and freshwater environments without going far from their village.

The two Barrancoid artificial mounds of Buckleburg-1 and -2 were constructed on the banks of the same river about 2 kilometers apart (Versteeg 1985). This likely represents a schema of settlements planned along a river where the principal site is found close to the coast and the secondary site is established slightly more to the south. It is easy to access one mound from the other, either by boat or on foot.

Hertenrits, in western Suriname, is the most obvious example of carefully defined boundaries. Stereoscopic interpretation of this zone indicates precise management and organization of the territory (Figure 54). The Arauquinoid site of Hertenrits is the largest of the known mounds, covering close to 4 hectares. Boomert (1980b) estimates that its construction would have required the equivalent of 14,000 loads from a medium-size truck. The mound is surrounded by a depression 20 to 100 meters wide, created to extract the clay from which the mound was built. The perimeter is crossed by five canals with traces of moorings, indicating places used as docks (Geijskes 1964). Several channels from water runoff were modified, some perhaps even created by indigenous peoples. The channels that cross the area have often been straightened or lengthened to connect one to another or to the canals. A complex network of straight, shallow canals crosses the landscape.

A number of connections were thus created among the different mounds, the raised-field complexes, and the streams. During the rainy season, the zone was crossed in canoes, while pedestrian paths prevailed in the dry season. Raised-field complexes consisting of irregular groups of less than 20 elongated fields occupied the surrounding thalwegs. Two smaller satellite mounds were constructed opposite to one another around Hertenrits. Wageningen-1 is 4 kilometers to the southeast, and Wageningen-3 is 3 kilometers to the northwest. Canals radiating from Hertenrits lead directly to these secondary settlements. The distribution and the connecting features demonstrate that the mounds were contemporaneous. They were constructed around AD 650 and occupied for several centuries before their abandonment.

The inhabitants of Hertenrits, Wageningen-1, and Wageningen-3 constructed their settlements following a precise model. They carefully divided the territory to take advantage of the full range of resources their biotope offered.

Although many Arauquinoid sites have not yet been found on the Guianese coast, current archaeological mapping demonstrates a specific settlement pattern. Artificial clay mounds are found in the western part of the Arauquinoid territory, whereas *chenier* sites are located in the eastern part. Settlements established on *terra firme* are located at the level of the final rapids of the principal rivers, serving as a junction point to the upstream populations. Raised-field complexes are systematically associated with villages, except for the Kwatta culture of central Suriname, where the fertile sand and shell bars provided natural foundations for raised fields. Causeways and canals facilitated the movement of boats and foot traffic, particularly to reach the ocean shoreline. The builders of raised fields were the most skilled of the coastal populations at exploiting the wide range of marine resources. The primary centers are distinguished by their size and probably served unique functions as well.

Thus, the size and location of sites demonstrates a hierarchy: a major center is generally established to the west of a major river, while smaller satellite sites spread to the west along sandbars. The principal centers of Sable Blanc and Bois Diable in French Guiana measure more than 5 kilometers long. Their eastern ends border the river, and the marine coast is easily accessible to the north. This ideal position gives them access to resources from the marine, riverine, forest, and coastal zones. They served as primary settlements for the Arauquinoid populations, while other farmers inhabited smaller villages to the west, adjacent to their raised-field complexes. These villages are connected by long paths that run parallel to the shore.

Both primary and secondary sites were established along rivers, at the level of the final rapids where the tidal zone begins, sometimes 50 kilometers from the ocean. These villages served as intermediaries to the inland populations and resources. Through either direct procurement or exchange, they acquired and redistributed products that were otherwise unavailable to the coastal communities. Suitable stone for tools is an excellent example of a necessary resource that is scarce on the coast. An important system of exchange of lithic tools from the interior to the coast of French Guiana has been demonstrated for the pre-Columbian period (Rostain 1994a, 2006b). Along the Middle Approuague, at Saut Mapaou, a group of 43 schist bifaces were discovered scattered over 20 square meters in the riverbed at the foot of the rapids. They consist of 24 direct percussion preformed blades, 14 partially polished pieces, and 5 finished polished blades. They probably represent the remains of the cargo carried in a canoe that sank while navigating the rapids, which posed a dangerous passage even for experienced navigators. The abundance of schist around and upstream from Saut Mapaou, along with its absence farther downstream and along the coast, indicates it is logical to assume the canoe was coming downstream when the accident occurred. Two hypotheses emerge to explain the expedition. The first is that coastal inhabitants traveled inland to procure lithic resources directly at their source and capsized on the return trip. The second possibility is that a group of specialists traveled downstream to exchange their products with the coastal populations. In each case, the blanks would have been created at the site of the raw material extraction, allowing for the selection of the best-quality specimens and eliminating stone that contained faults, thus lightening the load for transport. Furthermore, it appears that performing and polishing of certain pieces had already begun at the extraction site and was perhaps continued during the subsequent travel.

Around AD 1000 in Suriname, groups occupying the Brownsberg-Afobaka Massif of the interior exploited quarries and specialized in the production of stone tools, which they supplied to the coastal communities (Boomert and Kroonenberg 1977). On the central coast, groups of the Kwatta culture exchanged goods for these lithic artifacts and then traded them in turn to their eastern and western neighbors. The trade of *muiraquitãs* fits into this type of network. In the marshes of the western coast of Suriname, the inhabitants of artificial mounds of the Hertenrits culture had to travel 225 kilometers up the Nickerie River to procure the norite and granulite necessary to produce their groundstone and hammerstones. A myth of the Lokono of Pomeroon, Suriname, tells of the long and perilous voyage that was necessary to obtain the stone axes from the *hommes de pierre* (stone men) of the interior (Boomert and Kroonenberg 1977).

Obtaining stone was sometimes difficult because the selection of suitable source materials required special knowledge; thus, certain groups became experts in their procurement. The Kali'na of the coast of Suriname navigated along the Corentijn River as far as the Sipu Cove of southern Guyana to obtain their red burnishing stones (perhaps jasper) (Ahlbrinck 1931). Similarly, the Waiwái chose to walk for three days to reach the sources of metamorphic sedimentary rocks best suited for making the teeth for manioc graters, rather than exploiting the fluvial granitic outcrops located close to their villages. These stone teeth were so valued that when the wooden handle began to rot, the Wapisiana saved the stone pieces for a new grater (Yde 1965). In the Upper Mazurini of Guyana, some groups reserved exclusive rights to the exploitation of natural resources in their territory (Butt Colson 1973). Farther away in the Amazon, along the Upper Xingu, until the end of the nineteenth century, the Trumaí and Suyá held the monopoly on the use of stone quarries and the manufacture of stone tools (Lévi-Strauss 1948).

Amazonian ethnography has frequently shown that groups tend to specialize in the extraction and diffusion of specific raw materials, in hunting or fishing, or in the manufacture of specific objects. In Guiana, each tribe was known for the products that they traded. For example, the different Karib groups of the central plateau produced (and, in some cases, continue to produce) complementary goods that were mutually exchanged in a trading network. Historical archives indicate that specialized production among indigenous groups was much more common in the past than it is today.

In the Upper Camopi of French Guiana, noting that the Kaikusiána raised hunting dogs and that the Way, Piriu, and the Arouane produced and sold manioc graters fashioned with flakes inserted in a wooden board, Moreton de Chabrillan (1743) wrote that "Each of these nations has its object of trade and its particular manufacture, without any attempt to enter into competition with another. This serves to bind them together, requiring trade and essential contact among them." On the coast, the people of Cap du Nord are considered the best navigators and builders of canoes, which "are large, solidly constructed, and withstand perfectly the heavy seas" (Lescallier 1799), and "among the nations which have the greatest need, those of the Maillés excel at the making of Canots" (Préfontaine 1763). The Arikare of the central coast of Amapá intensively hunted manatees, which they sold to colonists (Barre 1666; Biet 1664). The Kali'na seem to have been great producers of cotton hammocks and traded not only with Europeans, but also with other indigenous groups. For example, they exchanged their wares for the "green stones" of the Arikare (Biet 1664). It is likely that some groups, such as the Maraon and the Itutan of Amapá, renowned as "excellent hunters and archers" (Préfontaine 1763), developed their skills in hunting and fishing to the detriment of other abilities, requiring them to exchange their game for other products that they required.

Even in the nineteenth and early twentieth centuries, Walter Roth (1924:635–6) remarked that each nation has its own specialized production and a reputation for its trade goods. He lists several examples:

- Otomac women were known for their pottery.
- The Makusi for their curare poison.
- The Maionkong and the Taruma for their manioc graters and their hunting dogs.
- The Warão for their shells.
- The Waiwái for their *tucum* and *kuraua* fibers.

- The Guinau for their hammocks, the calabash scrapers, their aprons, their belts of human hair and their featherworking.
- The inhabitants of the Oyapock River for their groundstone and their stone "spleene."

No product was considered useless because it could always be exchanged at some point. Among the trade goods, curare poison represented a highly desirable luxury item in western Guiana. At the end of the eighteenth century, along the Orinoco, only the Caberre were still producing it, to great profit for them, particularly at the large annual market that took place downstream from their community (Gumilla [1745] 1963). In the Upper Amazon, curare was only produced by the groups who lived beyond the rapids of the northern tributaries, particularly the Rio Negro and the Japurá.

The inhabitants of the Uaupès River exchanged stone flakes for making the teeth of manioc graters all along the Upper Amazon. In Guiana at the beginning of the twentieth century, the Makusi of the Rupununi River exchanged a minimum of one bottle of gunpowder in return for graters from the Arekuna, who had themselves procured the graters from the Maiongkong. The Makusi then exported the graters they received to Suriname through Karib groups. The Wapisiana obtained them from the Taruma. In sum, the Maiongkong organized an intense trade of manioc graters that extended from the Orinoco to the Rio Branco (Roth 1924).

On the Orinoco, during the Conquest, farmers controlled and cultivated the banks while fishers occupied those parts of the river best for fishing and foragers exploited the interfluvial zones. In addition to their agricultural products, farmers traded ceramics to nonagricultural groups in return for game, fish, and raw materials (Morey 1975).

On the alluvial plains of the Amazon Basin, along the eastern flank of Guiana, the first explorers discovered complex societies and specialized tribes who depended on each other for primary products:

- Groups of fishers accumulated stocks of dry fish, which they exchanged with the inland populations.
- The Caripuna and the Zurina sculpted wooden benches for trade.
- The Carabayana extracted salt from vegetal ash.
- The Aruaqui, the Tapajo, and the Trombeta exchanged their ceramics for cotton, cordage, tobacco, and maize.
- The Aiçuares also produced ceramics and cultivated cotton for other groups.

Some were specialized as traders and served as intermediaries in exchanges (Nordenskiöld 1930; Porro 1985). In the Amazon Basin, for almost three millennia before the Common Era, there existed long-distance trade, with some networks extending as much as 2,000 kilometers (Lathrap 1973).

Coastal Arauquinoid villages produced not only crops from raised fields, but also a variety of marine and other coastal resources. They were thus able to trade products that were lacking in the populations of the interior for resources absent from the coastal plain. Furthermore, certain groups were specialists in the manufacture of valued goods, such as the green stone *muiraqutãs* or ceramic griddles. The Arauquinoid world of Guiana developed in a sociopolitical space where goods circulated among different specialized groups. A theoretical model of management and hierarchy of this territory can be proposed, taking into account the different elements that made up Arauquinoid society (Figure 57). The spatial organization of the Arauquinoid villages as well as the specialization of different communities could therefore indicate a society characterized by a hierarchical system.

MAIZE VERSUS MANIOC

What crops were cultivated on raised fields? Many simplistic claims for the primacy of manioc in the diet of pre-Columbian peoples have been made, based primarily on the diet of modern Amazonian groups. However, this assumption ignores the profound changes that have taken place in the indigenous world since European contact, as well as the remarkable cultural diversity that characterized the ancient populations of the Amazon. The "manioc civilization" was perhaps previously less widespread than it is today in the region.

Raised-field techniques appear well adapted to the cultivation of tubers, but they are also appropriate for other plants. Bitter manioc (*Manihot esculenta*), maize (*Zea mays*), yams (*Dioscorea* sp.), and sweet potatoes (*Ipomoea batatas*) could also have been principal crops. Manioc tubers, when mature, penetrate 25 to 30 centimeters below the ground surface. Cultivated on mounds more than 30 centimeters high, they never reach the level of the matrix, where they would be at risk of rotting. However, their roots always extend to the level of the water, located 30 centimeters below the matrix during the dry season. The construction of mounds is thus an effective agricultural response to the problems posed by the seasonally flooded savannas.

In the absence of pollen data, the study of the artifacts associated with agriculture that have been recovered in the habitation sites (ceramic griddles, graters or mortars, metates, and stone pestles) provides useful data. Arauquinoid sites on the coast of Guiana contain more groundstones than other sites in the interior or on the eastern part of the coast. In contrast, graters or grater teeth, generally associated with the processing of bitter manioc, are rare. But not all methods of bitter manioc

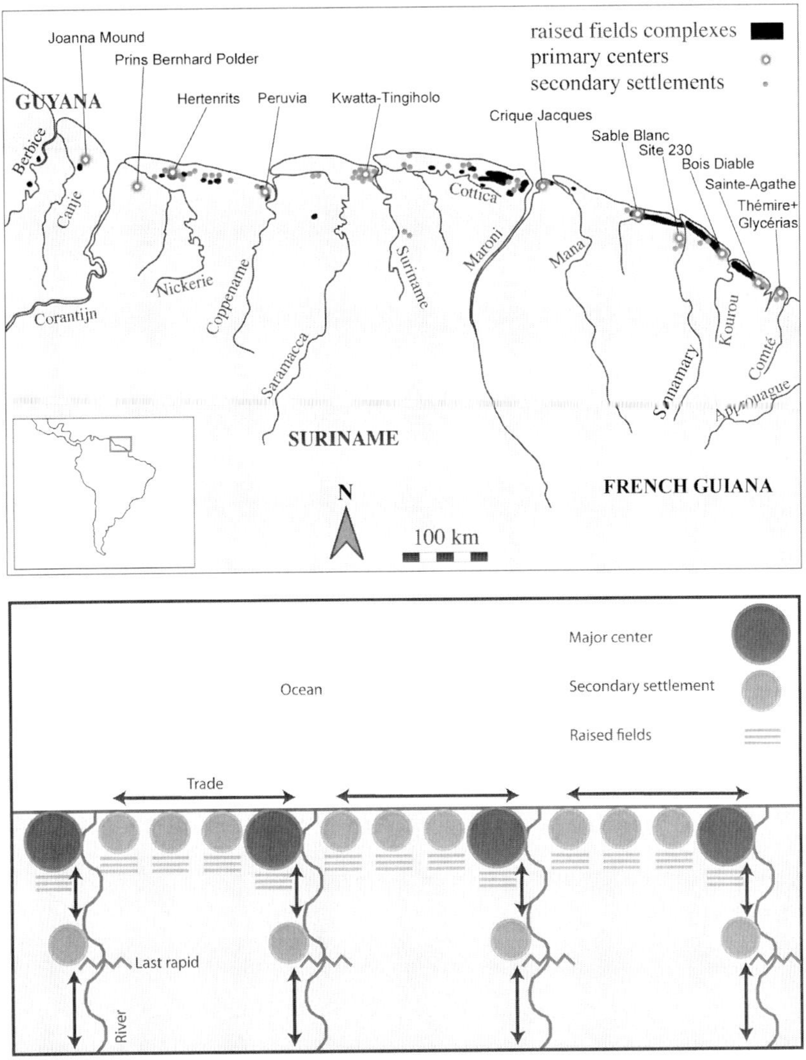

Figure 57. Top: primary and secondary Arauquinoid sites of Guiana. Bottom: theoretical model of the Arauquinoid territory organization (drawing by Stéphen Rostain).

preparation require grating. The Palikur of the Lower Oyapock macerate the tubers in the marshland (Doyle McKey, personal communication, 2010). This difference in the proportion of stone cooking tools could represent the preferred consumption of ground plants, such as maize. The ceramic griddles provide little proof either way, because they can be used for cooking cakes made from either manioc or maize.

Pollen analysis of samples from the Hertenrits raised fields on the coast of Suriname and from Ventosidad along the Middle Orinoco (Zucchi and Denevan 1979) does not reveal evidence of domesticated plants. This is likely due to the significant leaching of soils that is typical in tropical environments and to the decomposition, making pollen recovery difficult.

How then can we identify the plants that were cultivated on raised fields? The high temperatures and humidity of tropical regions, as well as their acidic soils, do not favor the preservation of macrobotanical remains (e.g., seeds, tubers, etc.). However, these environments tend to preserve a unique class of botanical microfossils, called phytoliths, or "plant stones," that can provide crucial information. These minuscule particles of silica are produced by a number of plants, and their morphology can often be used to identify the taxon (family, genus, and in some cases, species) or the group of plant (herbaceous dicotyledons, woody plants). Some plants that do not produce pollen diagnostic of their genus or species (such as numerous grasses and sedges) nonetheless produce distinct phytoliths. Phytoliths provide valuable information for archaeological and paleoenvironmental studies (Fisher et al. 1995; Kelly et al. 1998; Madella et al. 2002; Mercader et al. 2000; Pearsall and Piperno 1993; Piperno 1991, 1995; Vernet 1980). Combined pollen and phytolith analyses often provide complementary data. The advantage of phytoliths over pollen is that phytoliths are heavy particles that are not as easily displaced in the soil matrix. Pollen is generally dispersed into the air before becoming incorporated in soils or sediments, whereas phytoliths are contained in the plant tissue and are only released when the tissue decomposes on or in the ground. At the sites studied in French Guiana that are dominated by grasses and sedges, phytoliths provide a great deal of information regarding both the natural vegetation as well as the plants cultivated on the raised fields.

Another type of plant microfossil that has proven to be unexpectedly useful to archaeobotanists is the starch grain (Loy 1994; Piperno and Holst 1998). Although less resistant to degradation than phytoliths, starch grains nevertheless are preserved in certain microenvironments. Starch grains can remain protected for centuries, or even millennia, in the crevices of tools and ceramics (for example, the griddles or the vessels used to prepare and store food). Their structure is often diagnostic of genus or species, which permits them to contribute data regarding the plants cultivated and consumed that cannot be determined from phytolith analysis alone.

During fieldwork in 2005, I collected sediment samples for archaeobotanical analysis from four raised-field complexes: Maillard savanna, Carrefour Maillard, Grand Macoua savanna, and Piliwa. The samples were analyzed by archaeobotanist Magali Chacornac-Rault, who recently finished her doctoral thesis on the island of Java. These are the first data to directly reveal the plants that were likely cultivated on mounds in Guiana (Charcornac-Rault and Rostain, in press). As the sediments contained few phytoliths, the study relied principally on pollen analysis. The cultivated plants identified include sweet potato (*Ipomoea*), maize (*Zea mays*), malanga, (also called elephant ear; *Xanthosoma*), banana trees (*Musa*), taro (*Colocasia esculenta*), and sugar cane (*Saccharum officinale*). These results must be considered with caution because it is possible that the pollen recovered represents contamination by plants that were imported during the colonial period. Taro is only detected in French Guiana beginning in the eighteenth century, and sugar cane was not introduced until the seventeenth century. Indeed, although pollen can sometimes be more precisely identified than phytoliths, it also can diffuse more widely via air currents, potentially allowing pollen from recently cultivated specimens to contaminate the mounds. Furthermore, manioc pollen has only been recovered from the Grand Macoua savanna; a wild relative of manioc grows there whose pollen is indistinguishable from that of manioc. There is thus no evidence that the pollen recovered comes from species domesticated by indigenous peoples. Further study is needed to resolve the issue.

Between 2007 and 2011, as part of the "Archaeology and Ecology of the Coastal Savannas of Guiana" project, archaeobotanists José Iriarte (University of Exeter, United Kingdom) and Irène Holst (Smithsonian Tropical Research Institue, Panama) focused on the plants cultivated and consumed by Arauquinoid populations who built raised fields (Iriarte et al. 2010; McKey et al. 2010). They determined that the phytoliths recovered from the raised fields were derived from wild plants and maize. The domesticated species present in all samples was maize (*Zea mays* L.), with phytoliths derived from the rachis, the stem, and the leaves. The cultivation of species such as manioc or sweet potato cannot be verified because these plants do not produce diagnostic phytoliths. Squash (*Cucurbita*) was also identified from its distinctive phytoliths in the Piliwa complex in extreme western French Guiana, but maize was dominant everywhere. Furthermore, fragments of ceramic griddles were analyzed from the excavations at Sable Blanc, west of Iracoubo; maize and manioc (*Manihot esculenta* Crantz) starch grains were identified on their surfaces. The first two plants were cooked, while the latter was likely dried. Thus, there appears to be a correspondence between the plants cultivated on the raised fields and those consumed in Arauquinoid villages.

The comparison of different data permit us to propose that maize, squash, sweet potato, arrowleaf, and perhaps manioc were cultivated on raised fields during the pre-Columbian period.

The archaeobotanical results obtained in French Guiana correspond to those derived from Arauquinoid sites of the Middle Orinoco, where demographic growth around AD 800 is associated with the replacement of manioc by maize (Roosevelt 1980). Several studies have shown that maize and manioc were the principal cultigens of raised fields in South America (Darch 1983; Denevan et al. 1987; Spencer et al. 1994; Zucchi and Denevan 1979). Father Juan Gumilla's text ([1745] 1963) mentions primarily bitter manioc and maize, which were associated with other plants of lesser importance. This diversity of cultivated species in the same place recalls the multiplicity of plants grown in modern indigenous slash-and-burn fields (Grenand 1979, 1980, 1981). The Achuar of Ecuador cultivate a total of 100 different species, more than half of which can be found at any given time in the same field (Descola 1986). Although maize is less commonly grown today by indigenous groups in Guiana, it was still part of their diet during the colonial period: "their women attended to the cultivation of the fields, sowing their maize, and planting sweet potatoes and manioc" (Coréal 1722).

The phytoliths and starch grains that we have recovered in our study sites, and are now analyzing, show that even if manioc was part of the diet of raised-field farmers, it was clearly not the only cultivated plant. Yet when Europeans arrived in Guiana, the diet of the indigenous people was strongly oriented toward manioc, which has continued into the modern day. Archaeobotanical studies have permitted us to determine that significant changes in diet occurred just before the colonial period.

Pre-Columbian Demography

Throughout the world, agriculture in wetland zones (floating gardens, *chinampas*, raised fields, etc.) is always associated with a dense population (Denevan and Turner II 1974; Mollard and Walter 2008). Indeed, it is necessary to have a substantial labor force to build and maintain these systems. On the other hand, these techniques provided food for large communities and even permitted the accumulation of surplus for storage or exchange. The construction of these systems also often implies the presence of specialists to organize and manage the distribution and construction of the fields. Therefore, the building and the use of raised fields could indicate a specialization of labor and, perhaps, some degree of hierarchy.

Ethnohistorical data, as well as archaeological research carried out in Latin America (and elsewhere in the world), shows that marshes are generally cultivated during periods of strong population growth, where shifting cultivation in the uplands provides insufficient yields (Denevan and Turner II 1974). Thus, the large number of Creole slash-and-burn fields observed on aerial photographs demonstrates how a small group of cultivators could quickly occupy and exhaust the nonflooded areas of the coastal plain. Similarly, the exhaustion of soils on the hills of Île de Cayenne and repeated invasions by leafcutter ants (*Atta* sp.) at the end of the eighteenth century necessitated a search for new farmland in French Guiana (Hurault 1989; Le Roux 1992). Colonists turned to creating polders in the lowlands, which they had previously avoided.

It is clearly difficult to estimate the size of the populations that inhabited the coastal plain of Guiana and cultivated with raised fields, because the construction and cultivation of these fields stretched over several centuries. Nevertheless, the number and the physical extent of raised fields, as well as the simple fact that such difficult land was selected for cultivation, probably due to the lack of sufficient highland fields, indirectly attests to the high density of human population along the coast. Studies of aerial photographs demonstrate that the entire surface of some areas was modified by the construction of raised fields, whereas in others, only areas that flanked sandbars were characterized by mounds. These differences in the density of drained fields may correspond to the density of communities present. Furthermore, the sandbars may have offered a supplementary space for agriculture, although probably with lower fertility.

The construction, maintenance, and cultivation of raised fields required well-organized communal labor. Because the hydrological works involved require careful planning, these systems were likely accomplished by communities that had grown accustomed to such work, and were possibly supervised by a specialist.

The work was accomplished by the inhabitants of the villages on the *cheniers* and artificial mounds. To better estimate the population density of the Arauquinoid mound builders, it is important to consider the demography of the modern Amazon. Although the European conquest and five centuries of colonization have drastically altered the indigenous world, modern farming populations still remain the best references for such exercises. Although modern statistics cannot be simply transposed to pre-Columbian contexts, they provide a useful comparative scale.

There is a large variation between the average extent of cultivated zones per person among modern Amazonian groups: for the Yanoama (Niyayoba-Teri), 405 square meters; the Yanoama (Jorocoba-Teri), 607 square meters; the Cubeo, 810 square meters; the central Yanomami, 900 square meters; the Achuar, 1,371 square meters; the Siona-Secoya, 1,970 square meters; and the Kuikuru, 2,632 square meters (Descola 1986, synthesizing several references). In contrast, in French Guiana, the average cultivated area by swidden field by family varies only slightly among groups: for the Kali'na of the Lower Mana, 3,300 square meters; the Wayana of the Upper Maroni, 4,200 square meters; the Wayampi of the Upper Oyapock, 5,100 square meters; the Palikur of the Lower Oyapock, 7,500 square meters; and the Boni (*Noirs-Marrons*) of the Lower Maroni, 8,100 square meters (Gely 1984). This means that a slash-and-burn field must be at least 5,000 to 8,000 square meters (0.5–0.8 hectares) for a family to be self-sufficient for a period of one year. An even larger size would be necessary for the production of a surplus harvest (Gely 1984).

However, the productivity of raised fields is significantly superior to that of swidden agriculture. Along the Middle Orinoco, the Karinya practiced until recently highly productive agriculture on drained fields that required only 0.39 hectares per family (Denevan and Schwerin 1978). Furthermore, in the state of Barinas in the Venezuelan *llanos*, modern farmers use pre-Columbian raised fields to grow maize, beans, and manioc, as well as other crops, achieving two harvests per year (Spencer et al. 1994). The improved productivity of raised fields over swidden agriculture is perhaps the reason that the raised fields of French Guiana were built at two distinct elevations; the mounds in the more humid zones may have been cultivated and harvested in the dry season, while those found at higher elevations were perhaps used in the rainy season.

In several cases, the extent of raised-field complexes is correlated with the size of the associated residential system. The Colline Sable *chenier* settlement, which stretches 3 kilometers in length, is bordered by numerous mound complexes in the southern marsh. The relation is clear between artificial residential mounds and elongated raised-field complexes in southern Suriname. The small Burnside Mound is accompanied by a modest raised-field complex. Larger complexes, covering 53 hectares, are associated with the larger mounds of Buckleburg-1 and -2, while the largest mound at Hertenrits is surrounded by a 160-hectare zone of raised fields (Boomert 1980b).

Based on the demography of raised-field farmers in tropical zones, and the productivity of this type of agriculture, I propose a density of 50 to 100 inhabitants per square kilometer in the Arauquinoid territory (Rostain 2008b). But this estimate must be adjusted for individual cases. The Piliwa complex, in the Lower Mana in French Guiana, extends into a flooded depression 1,200 meters by 50 to 150 meters. The farmable surface of the mounds reaches 5 hectares, which could have served the needs of several families. Newer, more precise estimates have been made in the "Archaeology and Ecology of the Coastal Savannas of Guiana" project (Iriarte et al. 2010). José Iriarte used data obtained from various agricultural experiments on raised fields to estimate the population that could have been supported by the production of maize at the Grand Macoua savanna complex. In this large savanna, the zone modified by raised fields is 167 hectares, almost the entire surface. The farmable area only represents 45 percent of the modified zone, or 75 hectares. The calculations assume the consumption of 160 kilograms of maize per person annually (with the remainder of the diet fulfilled with other resources) (Sanders et al. 1979). Based on the lowest estimates of maize productivity, 2 $t.ha^{-1}$ (tons by hectare) (Arce 1993), with a single annual harvest and assuming that only 25 percent of the fields are being used at one time (the others in fallow), the production of maize would have been able to feed 234 people in the Grand Macoua savanna. A moderate estimate, assuming that 75 percent of the fields are in cultivation and productivity is 3.725 $t.ha^{-1}$ (Sanders et al. 1979), would have been able to support 1,310 people. If we apply the maximum productivity of 5.78 $t.ha^{-1}$ (Muse and Quintero 1987), 100 percent of fields in cultivation, and two annual harvests, the estimated population that could be supported rises to 5,400 people. I think that it is reasonable to rely on the moderate estimate. Newer estimates based on additional criteria support those proposed here. This concordance of results indicates their plausibility. Thus, the 3,000 hectares of raised fields in French Guiana would have been able to support between 100,000 and 300,000 people. However, this estimate should be lower because it is unlikely that all of the complexes were cultivated simultaneously.

Although it is almost certain that the coast of Guiana was densely settled by Arauquinoid communities around AD 1000, it is difficult to make precise demographic estimates. The agricultural capacity and surface area covered by raised fields suggest that a large farming population inhabited the coast of Guiana. Comparing the productivity of swidden agriculture with that of modern and experimental raised fields, and then extrapolating, produces a figure of 50 to 100 inhabitants per square kilometer. This estimate most likely does not apply to the entire coast because it is unlikely that it was completely and uniformly inhabited. Archaeological dating suggests that Arauquinoid settlement extended progressively through time toward the east.

In 1492, with the arrival of the Europeans, the population of Amazonia was estimated to be between 5 and 7 million inhabitants (Denevan 1992a). Two centuries later, more than 80 percent had disappeared, and it is estimated that only 153,000 indigenous people remained in 1730. Nevertheless, this population was not evenly distributed across the lowlands (Bush and Silman 2007). Although it is unlikely that the interfluvial zones were ever densely settled, other regions of Amazonia, such as the Amazon and Orinoco Basins, the Andean piedmont, and coast of Guiana were home to large communities at different periods in time. Furthermore, it seems that the demographic declines began well before the arrival of the Europeans, around AD 1200, perhaps due to climate change (Meggers 2007; Rostain 1994a).

Until recently, anthropologists proposed a density of between 1 and 3 persons per square kilometer in the Amazon for both the colonial and pre-Colombian periods. Ongoing archaeological research on raised fields permits us to drastically revise those figures. In the San Jorge Valley of Colombia, an area of 1,440 hectares of raised fields and 400 residential platforms is estimated to have been home to a population of 2,400. More precisely, based on the area of the residential platforms, a population density of 160 persons per square kilometer is assumed for the period between ad 200 and 900 (Plazas and Falchetti de Saenz 1987). On Marajó Island, at the mouth of the Amazon, Anna Roosevelt (1991) proposes 5 to 10 inhabitants per square kilometer, Michael Heckenberger et al. (2003) propose 6 to 12.5 inhabitants per square kilometer on the Upper Xingu, and William Denevan (1992a) proposes 15 habitants per square kilometer along the Amazon River. Whatever their circumstances, raised-field agriculture is not a technique used by small tribes but by quite large groups. It is likely that the tribes that created and used the pre-Columbian raised-field sites in the Americas were characterized by social complexity beyond that of simple family groups. They were likely stratified communities (Erickson 2006a; Scarborough and Lucero 2010). In any case, it is probable that some of the sculptors of the lan landscape developed a hierarchical social structure.

Chapter 5

"500 Years of Solitude"

What have become of Guiana's savannas since the Conquest? It is difficult to specify when indigenous peoples abandoned raised-field farming practices. It is likely that this had already occurred in Guiana by the time of the Europeans' arrival. However, there are a few exceptional mentions of this practice in the region in historical archives (Figure 58). Raised fields were still cultivated in the Venezuelan savannas from the sixteenth to the eighteenth centuries (Gumilla [1745] 1963). Some rare cases are attested to in the following periods, but the drop in population occasioned by disease epidemics limited the size of indigenous groups, who no longer had reasons or means to continue to produce and maintain these works.

Although some archaeologists were enthusiastically convinced by the discovery of raised fields in French Guiana 20 years ago, others evinced skepticism. Similar debates had arisen among scientists in previous years regarding Mayan raised fields, and whether they were of pre-Columbian or colonial origin. For example, the canals in Campeche, Mexico, initially attributed to the pre-Columbian Mayans, were later found to date to the nineteenth century (Millet Cámara 1984). As archaeological fieldwork and data increased, the antiquity of landscape modifications was demonstrated in several areas of Latin America. In Mesoamerica, the first raised fields were constructed perhaps beginning in the Middle Preclassic (1000–300 BC), and reached their height during the Classic period (AD 200–1000) (Adams 1982; Darch 1983; Denevan et al. 1987; Hammond 1986; Harrison and Turner II 1978; Turner and Harrison 1981; Zola Baez 1986). Similarly, increasingly precise analyses of the earthworks in South America have revealed a similar antiquity and an intensification of this agricultural practice during the first few centuries of the Common Era.

In French Guiana, some scientists assumed that raised fields had been built by slaves or prison labor. Indigenous technology and social organization were again considered a priori to be inferior. This assumption recalls reactions to the first archaeological discoveries of the eighteenth century. When Captain Pinto da Gaya found the Montagne d'Argent Petroglyph of French Guiana in 1728, it was interpreted as a boundary marker representing the seal of the king of Spain, carved by Europeans

Figure 58. "*La mañiere et facon de jardiner et planter des Yndiens*" (the way to do gardening and to plant of the Indians) on an agricultural bed at the end of the sixteenth century in the Caribbean (Drake 1996).

at the beginning of the colonial period (Hartt 1871a). It was inconceivable that indigenous groups could have engraved those rocks. I thus turn to colonial history to verify whether raised fields were constructed when the colonists had access to significant forced labor. No mention of raised-field construction is found in the texts. The colonists, few of whom had rural backgrounds, were most often overwhelmed by this heretofore-unknown landscape and tropical climate. In this depopulated land, they attempted to farm without an understanding of the poor soils and the biseasonality of the climate, which was nothing like the four seasons of Europe. Many colonists, lured by the apparent lushness of the environment, learned the hard way that Guianese fertility was an illusion. Many then chose to imitate the indigenous peoples and adopt shifting agriculture in swidden fields. On the other hand, Europeans undertook large construction projects in the lowlands. Although it is evident that French prisoners had nothing to do with these structures (why would escaped prisoners build raised fields in the savannas of Suriname and Guyana?), slaves built immense polders, notably in Suriname and Guyana. The colonists had indeed carried out agricultural modifications in the lowlands. They participated in a remodeling and deliberate anthropization of the landscape, similar to that of the building of raised fields.

More recently, the technique of using raised agricultural beds on dry land spread along the coast of Guiana. The remains of these mounds and their associated canals can be found everywhere. In contrast to pre-Columbian sites, the Creole settlements of the nineteenth and twentieth centuries were dispersed, each independent of the others. They extend in chains along sandbars not far from colonial roads that run parallel to the shore. The absence of historical records or study of these recent occupations necessitates the use of archaeological research and interviews with persons old enough to recall these periods to document life on the savannas at this time. Recent archaeological survey has revealed previously unknown aspects of this society and its lowland settlements. Raised-field agriculture in the savannas has not totally disappeared today, but it is not practiced to anywhere near the extent it was in the pre-Columbian period. A few limited and short-lived attempts were made along the coast. These raised fields are smaller, and the complexes only cover small areas. They are also rarely built in wetland areas because the farmers preferred the dry summits of sandbars and *cheniers*. Recent immigrants from Haiti are the principal practitioners of this modern agriculture, introducing techniques that were developed initially on their island and, before that, in Africa.

What of the modern state of the pre-Columbian raised fields? Although the terrestial environment is relatively free from erosion, the tropical climate strongly affects the state of soils, particularly through leaching action. Nevertheless, modern human impact is the most destructive force in the region (intentional burning, road and building construction, mechanized agriculture, etc.). Modern humans' race to destroy the world has not spared the raised fields. If nothing is done to protect these ecofacts, our current memory of them will constitute the chronicle of a death foretold. The reasons for the erosion and destruction are such that it seems extraordinary that these remains have not already been completely erased. Some even appear to have been recently reconstructed and in good condition.

THE LAST INDIGENOUS RAISED FIELDS

Raised-field technology disappeared almost completely from the Amazon during the colonial period, most likely due to the decline in population. A few groups continued to practice this method of farming (in some cases, into the modern day). In Venezuela, the chroniclers describe agriculture on mounds. In French Guiana, archives from the eighteenth century indicate that this practice was still in use at the time. Pierre Barrère (1743:58) noted the cultivation of manioc through cuttings on "*grosses mottes*" (large mounds) in the lowlands: "In the flat lowlands, to prevent manioc to rot, they make big mounds, where they plant four wooden sticks."

Archival texts, as well as ethnohistory, mention that indigenous peoples had once cultivated drained fields, particularly in marshy areas. In the sixteenth century, with the arrival of the Spanish at Hispaniola, the northern plains and the flooded central valleys were intensively cultivated using this technique (Dreyfus 1981, 1993). In 1589, Juan de Castellanos noted "*camellones*" (elongated raised fields) and "*calzadas*" (roadways) in the Venezuelan *llanos*. In 1647, Fray Jacinto de Carjaval observed earthen structures (cited in Zucchi and Denevan 1979). In 1745, after observing indigenous cultivation of the marshes of the Venezuelan *llanos*, Father Juan Gumilla described with exceptional detail the tool used to build raised fields ("*macana*") and the method of fertilizing the mounds using vegetal debris:

> The barbarians who lived and still live in the shrubby savannas, not having the difficulties associated with trees and forests, obtain their fruits, although in less quantity, with less work; because with the club shovels of which I was speaking, they lift the earth (in humid places) from one side and then the other of the furrow, covering the straw and the hay with the earth extracted from one side and the other, and after they sow their maize, manioc and other root crops, and in all parts a large quantity of pepper . . . (Gumilla [1745] 1963:429–30).

In northwestern South America, several chroniclers have also noted select episodes of farming on mounds or elongated raised fields (Darch 1983).

The most precise data are ethnohistoric. During the seventeenth century, the Palikur still numbered 1,200 along the Uaça River, in northern Amapá. The size of this zone (150 square kilometers) did not allow for the use of shifting agriculture, as this technique would not produce sufficient quantities of food to feed the population. Raised fields were therefore constructed in the flooded savannas and cultivated for several years consecutively. With the demographic decline, and the corresponding decrease in overall food requirements, the mounds were reduced in size as either round mounds (*imukwi hipatip*) 80 centimeters in diameter and 30 to 40 centimeters high or elongated raised fields (*imukwi kiawimin*) 200 by 50 centimeters, surrounded by a ditch used for irrigation and drainage. This technique was principally reserved for farming bitter manioc, and to a lesser extent, yams (*Dioscorea trifida*). With increased demographic decline, they abandoned this method at the end of the nineteenth century and instead relied exclusively on shifting agriculture (Grenand 1981).

The development of the pre-Columbian cultures in Guiana and the origin of the modern groups were discussed in a previous chapter. The Palikur clans may have emerged from the fusion of local populations with the communities that came from the Lower Amazon and the coast of Guiana in the sixteenth and seventeenth centuries (Rostain 1994a, 1994b). Two groups appear to have influenced the Aristé culture, indigenous to the Lower Oyapock and northern Amapá, after the Conquest. One of these groups came from the left bank of the Lower Amazon and would have been associated with the Mazagão culture. The other, from the Arauquinoid tradition, would have come from the western coast of Guiana. The archives indicate that the indigenous peoples fled from Portuguese slave hunters in the Amazon estuary and from their Spanish counterparts in the Orinoco Delta. The people persecuted in the latter region fled mostly to the east, toward Amapá. This cultural mixture is particularly visible in the evolution of ceramic styles. The development of pre-Columbian ceramic styles at Amapá that has led to modern Palikur pottery attests to the emergence of exogenous styles in Late Aristé ceramic art. This style resulted from the arrival of refugees from the west and the south, shortly after the Conquest. The forms and the decoration departed progressively from the characteristics of Late Aristé pottery and began to approach those of the Mazagão and Arauquinoid styles. This double influence on the Aristé culture would have produced a new, hybrid style that would have eventually become an original complex in its own right. Thus, the colonial fusion of Aristé, Mazagão, and Arauquinoid tradition led to the genesis of the modern Palikur ceramic tradition. At the beginning of the twentieth century, Palikur pottery already possessed the same technical and artistic characteristics as it does today (Nimuendaju [1926] 2008; Rostain 1992). Neither significant indigenous migration nor external influence has altered this style for more than a century.

The evolution of Aristé ceramics during the colonial period accompanied the gradual abandonment of ancient funerary cremation practices, which dominated during the Late Aristé, as well as the adoption of secondary burials of defleshed skeletons, typical of the Mazagão and Arauquinoid cultures. This schema agrees with the Palikur origins story, as it is recorded in oral tradition and historical archives (Grenand and Grenand 1987). It is highly likely that the refugees from the western part of the coast of Guiana also brought with them new technologies, such as the construction of raised fields (Rostain 1994a). This would explain their usage by the Palikur during periods of population growth.

Raised fields have been almost completely abandoned today throughout South America, and were already rare in the colonial period. Nevertheless, a few examples of this practice, and related ones, still exist. In the Peruvian and Ecuadorian Andes, small, narrow, elongated raised fields are used, associated with depressions (*qochas*), principally for farming sweet potatoes (Denevan 1963; Denevan et al. 1987). Among the modern mixed populations of Paraiba state, in northeastern Brazil, elongated raised fields permit the cultivation of zones with low fertility and poor drainage (Prost 1967). In the lowlands, until recently, the Makushi of the Upper Branco of Brazil, the Karinya of the left bank of the Lower Orinoco, and the Guató of Mato Grosso also cultivated on raised fields (Denevan 1966, 2001; Reichel-Dolmatoff and Reichel-Dolmatoff 1974; Zucchi and Denevan 1979).

The Karinyas of the Venezuelan *llanos* do not construct raised fields. Rather, they use an ancient system to improve soil fertility by using canals to drain the *morichales*, formations of *moriche* palms (*Mauritia flexuosa*) with fertile but flooded soils. Local farmers design a large network of canals spaced 20 meters apart. These channels range from 10 to 100 meters long, 20 to 70 centimeters deep, and 70 to 100 centimeters wide. The parcels that are delimited by the canals are further divided by smaller secondary and tertiary ditches (Denevan and Schwerin 1978). This technique allows for year-round planting and harvesting (Denevan 2001). While it is true that the construction of these fields requires a significant investment of time and labor, the authors remark that they also allow for an intensive agriculture that is much more productive than slash-and-burn farming. Furthermore, once the fields are drained, they require very little work to maintain. As with raised fields, after their somewhat intensive initial construction, their maintenance is fairly simple; only periodic weeding is necessary. The canals must also be cleared annually prior to planting. Maximal fertility is achieved in two to four years, at which point the field can be farmed for many years (Denevan 2001). It is very probable that this type of drained agriculture was more widely practiced in the pre-Columbian period.

The knowledge of raised-field agriculture has almost completely disappeared from indigenous groups today. However, some anecdotal accounts exist to indicate that this practice persisted until fairly recently. A young Kali'na from Kourou still remembers his aunt cultivating manioc, bananas, and yams on small mounds (Schnier 2009). During low-altitude flights in 2008, I also observed mounds in a plot that had been recently cleared on the *chenier* of the Kali'na village of Awala.

Colonists and the Lowlands of Guiana

The first colonists fled from the lowlands and settled the uplands, convinced that the "fumes" and "noxious miasmas" of the stagnant waters provoked "putrid" fevers (Le Roux 1992): "The fatal vapors of this virgin land kill men who seek to open it carelessly" (Pitou 1805:223). The well-recorded history of Guiana does not indicate any construction or use of raised fields. Only during the period of slavery did Guiana posses a workforce suitable for undertaking earthworks on such a large scale. The only subsequent period where a comparable labor force would have been available was in French Guiana during the penal-colony period.

Thus, it is necessary to return to the debate over the fertility of Guianese soils. The first colonists of Guiana were ecstatic upon observing the exuberant vegetation, which they interpreted as evidence of exceptional abundance. This perception gave birth to a polemic that continues even today: the defenders of inexhaustible fertility versus the proponents of much more limited abundance, or even sterility of the soils. Despite the successive failures of agricultural attempts since the beginning of the colonial period, the assumed richness of Guianese land was the leitmotif of many explorers, administrators, scientists, and even agronomists. The reasons for this unwavering optimism are not always very obvious, often hiding obscure and somewhat disreputable motives where personal ambitions mixed with political motivations. Theoretical consideration won out regularly over the wisdom of lived experience, which was systematically denied or discredited. The recurring tragedy of Guiana consisted of new arrivals who, deluded by unrealistic and idealized schemes, ignored the skills, experiences, and the mistakes of their predecessors, only to repeat these errors again themselves. Earlier failures were systematically blamed on lack of technology. Recently arrived colonists boasted about finding the solutions. For five centuries now, the same ecological deceptions have thwarted the hopes and dreams of gullible innocents who launched themselves into rural adventures in Guiana (Turenne 2008). We note, for example, the disaster of "the farming utopia" of New Angoulême on the Mana River, in French Guiana, at the beginning of the nineteenth century (Durand 2010).[1] The myth of great agricultural potential persists into the modern day. The French Green Plan of 1976 promised that "... on this fertile earth, we will plant manioc, bananas, rice, maize ..." (cited by Turenne 2008). Unfortunately, this Promised Land goes back on its vows more often than it honors them. Only through the use of large-scale earthworks can the Guianese lowlands become exploitable and sustainable.

The Polders

Worked mercilessly, slave labor allowed rich colonists to undertake ambitious projects. Nevertheless, agriculture was limited at the beginning of the colonial period to opening clearings in the "highland" forests. Until the eighteenth century, only the uplands in French Guiana were cultivated by slash-and-burn because "its first Colonists, frightened by the great abundance of rains and by the state of the lowlands, always very humid and sometimes completely flooded, decided to go to the mountains or highlands, a shelter against this periodic flooding" (Lescallier 1799:53–4) and "without speaking of the flooded savannas, which did not appear suitable for agriculture" (Lescallier 1799:51). A few years previously, in 1761, the engineer Beteow located "beautiful and very fertile plains that the new French colony must inhabit" between the Kourou and Maroni Rivers on the map of equinoctial France. He thus fed the dreams of fertility for the metropolitan intelligentsia who had never actually confronted this terrain.

An alternative arrived during the mid-seventeenth century, first in Suriname, with the construction of polders. The word "polder," literally "confined earth" in Dutch, was used for the first time in 1219, in a charter of Middelbourg of the Zeeland province in the Netherlands (which means "Earth of the sea"). Initially, the marshes were drained for cultivation during the period of low waters; later, the technique was perfected. The polders are flooded parcels that are surrounded by dikes that serve to drain them. Technically, these are reclaimed areas that lie below sea level. By extension, the term refers to the very humid lowlands, confined or drained. The polders of Guiana are located in the low recent plain that is alternatively flooded by the sea and by fresh or brackish water. Thus, the limit of salinity is in constant motion depending on the tides and the seasons. Using their expertise for building dykes developed in their home country, the Dutch began to "polderize" 1- to 2-kilometer-wide swathes of the alluvial soils along the banks of the downstream portions of the Essequibo, Demerara (Suriname), and Berbice (Guyana). Nevertheless, the poor circulation of the water allowed for the concentration of salt, which caused the soils to become sterile. Thus, many of these polders were abandoned, and beginning in the eighteenth century, farmers turned to the estuaries; later, in the nineteenth century, they moved to the coastal plain (Giacottino 1984).

The Dutch constructed polders over thousands of hectares in Suriname (Figure 59). The English followed suit in Guyana, taking advantage of the abandoned Dutch construction. But, again, the desalinization of the soils was prevented by the deficient natural drainage. It was therefore necessary to construct structures to alleviate this problem. Dikes protected the parcels from the strong tides, and large canals evacuated excess water. Irrigation was provided through a system of drainage that drew water from the rivers or from reservoirs excavated to the south of the polders. The technique consisted of first excavating the enclosing canal, with the canal spoils serving to elevate the dikes. Then a drainage ditch, controlled by a lock, led back to the ocean. Finally, secondary canals separated the planting beds within the complex. It was also necessary to flood the polder every 30 to 40 years to clean out the insects and regenerate its fertility (Le Roux 1992).

Although there were only 23 plantations in Suriname in 1668, by 1800 there were 641 active along the length of the Para River and in the estuaries of the Suriname and Commewijne (The National Planning Office of Suriname and Regional Development and Physical Planning Department [NPOS and RDPPD] 1988). Sugar cane was the

Figure 59. Polders east of Paramaribo and north of Cottica River, Suriname (photo by Stéphen Rostain).

principal crop, transported on barges along the canals to the sugar factory to be processed. It is clear that, under the Ancien Regime, it was the Dutch who most skillfully mastered the technology and production of Guianese polders. The official production of Suriname represented 14 times the average value of that of French Guiana in 1778 (Souty 1982).

In French Guiana in 1764, the engineer Samuel Guisan, trained in Suriname, taught the polder method to the French and 24 years later published a detailed manual on the technique (Guisan 1788). In the same year, Claude de Macaye constructed 20 hectares of polders in the Fonds de Rémire of the Île de Cayenne to cultivate coffee (Le Roux 1992). The colonists began to very gradually abandon the highlands, which had been exhausted, to cultivate the lowlands, but the French Revolution soon halted that movement. A new phase of polder construction resumed from 1815 to 1848. Some polders were built along the coast of the Île de Cayenne up to the Oyapock, particularly on the right bank of the Mahury and the downstream Approuague. A long canal also was excavated in Torcy in 1821, through the marsh of the Lower Approuague to facilitate the transport of sugar cane. The polders were for the most part abandoned in French Guiana after 1848 (Giacottino 1984; Mam Lam Fouck 1999). In contrast, they were almost nonexistent west of the Île de Cayenne.

The polders are the most remarkable expression of the Cartesian mindset imposed on Amazonia by Europeans, quite distinct from the symbiotic attitude of the indigenous groups. The eastern coast of Suriname and the entire coast of Guyana were completely transformed by these modifications, which fundamentally altered the landscape of the lowlands. The coastal plain of Guyana and Suriname is almost totally covered by polders.

The construction of polders gradually decreased over the course of the nineteenth century with the abolition of slavery, but increased again in the twentieth century due to the availability of machines that replaced slave labor for draining large areas. At the beginning of this century, large polders were built on the eastern coast of Guiana between the Pomeroon and Corentijn Rivers. The western portion of the Surinamese coast, previously neglected, was subject to enormous earthworks beginning in 1949 with the Wageningen Project between the Nickerie River and the ocean. Indian and Indonesian immigrants introduced rice farming to Guiana, stimulating the colonial governments who envisioned being able to transform Guiana into a granary. Rice paddies stretched over 115,000 hectares in Guyana and 40,000 hectares in Suriname (Giacottino 1984). Despite this difference in total area farmed, the yields were greater in Suriname due to better techniques and management.

Today, few polders are active, and are essentially used for rice, then sugar cane, and finally, to a lesser extent, root crops and vegetables. In Guyana, for an area cultivated twice in 1980, the rice exports still brought in 3.5 times less revenue than those of sugar cane (Giacottino 1984). The most extensive polders are found on the western part of the coast of Guyana from the Nickerie district to western Suriname, as well as those of Mana to the western border of French Guiana. Nevertheless, the surface areas are significantly reduced. In Guyana, the surface area of the rice paddies has fallen from 100,000 hectares in 1964 to 36,000 hectares in 1988. In Suriname in the 1970s, only 3,500 hectares of polders produced rice in the Coronie district and along the western portion of the coast (NPOS and RDPPD 1988). In French Guiana, the rice polders of the Lower Mana extend over 5,500 hectares.

"Penal Agriculture" of French Guiana

From 1851, prisons were particularly prevalent along the western coast of French Guiana, and most particularly at Saint-Laurent-du-Maroni. Prison labor was mainly used for building colonial roads and for cutting trees. Few crops were grown, as most of the provisions came from France or Brazil. The archives and other historic texts do not mention large-scale prison-based farming. Instead, they focus on the difficult tasks of tree-cutting or road construction (which produced some 60 to 70 deaths each month). The convicts obviously carried out significant works, such as building the port facilities and the colonial roads or completely clearing Devil's Island (*Îles du Salut*) of trees.[2] Although no text mentions the construction of polders by convict crews, archaeology suggests otherwise.

During survey in the summer of 2003, we discovered a small polder in the savanna that borders the right bank of the Organabo River. A system of polders that measured 150 by 80 meters had been excavated in the middle of a group of pre-Columbian raised fields. After measurements and recording, we surveyed the forest where the remains of the habitation were located. These remains consisted of bottles and pottery sherds from the early twentieth century as well as foundation blocks made from consolidated lateritic rocks. The study of historical documents shows that these correspond to the supports for wooden posts in typical colonial houses. On the other hand, maps show no evidence of a civil operation in this location, nor had colonists advanced to western French Guiana before the nineteenth century, as their settlements did not extend beyond Sinnamary until much later. The texts do, however, indicate the existence of a small prison camp at Organabo. The spacing between the plots corresponds to those of prison barracks (Le Roux 2006:82). It is possible that a small group of prisoners lived here, farming the small polders for their basic subsistence. The most unusual observation is that this polder is located in the ancient coastal plain that remains unflooded year-round.

The convicts therefore did not construct large agricultural complexes or raised fields over large areas. I have nevertheless discovered an example of noteworthy agricultural constructions in an obscure source (Péan 1948). From 1920 to 1940, Charles Péan, officer of the Salvation Army, dedicated himself to the well-being of the convicts of French Guiana and after the elimination of the penal colony was involved in the management of those freed. He described a Salvation Army farm established in August of 1933 with approximately 20 recently freed men, apparently located on the Montravel hill, at Île de Cayenne. In the 1980s, while doing survey on the island, I discovered the remnants of large wells 3 to 4 meters in diameter that were likely remains of this farm. In six months, the freed prisoners had built fields, henhouses, and pigsties. They cultivated maize, bananas, some fruit trees, and "their agricultural beds regularly produced beautiful vegetables" (Péan 1948:195). He referred to large, low agricultural beds that are arranged in parallel and printed an exceptional photo of these structures (Figure 60).

Furthermore, although the prisoners occasionally cultivated some raised fields, the totality of the agricultural mounds on the savanna cannot be attributed to them. The deportees did sometimes cultivate raised fields, particularly on one parcel on the left bank of Passoura Creek, near Kourou, but these "convict gardens" were of modest dimensions (Marc Boyé and Jean-Jacques Rostan, personal communication, 1990).

Figure 60. Agricultural beds made by convicts in the Île de Cayenne, French Guiana (Péan 1948:208bis).

One can easily imagine that some prisoners from farming families had noticed the mounds in the marshes, and at times took up this technique of raised-field farming as it was practiced in France (Mollard and Walter 2008). In 1989, I interviewed several elderly Creoles from Kourou who had previously lived in the surrounding savannas before being displaced by the construction of the Centre Spatial Guyanais. None of them had any recollection of the mounds being farmed during the penal colony. Nevertheless, popular memory of the 1960s basked in the idealized memory of immense prairies that were cultivated by the penal colony. The reality is less spectacular. In 1944, barely 22 hectares were cultivated: 8.29 hectares of staple crops, 7.47 hectares of vegetables, 2.8 hectares for animal feed, and 2.76 hectares of bananas (Turenne 2008). Prison agriculture thus remains anecdotal.

Creole Agricultural Beds

Historians of French Guiana tend to dwell on the period of slavery and the penal colony, particularly to criticize them. Works on the nineteenth and twentieth centuries therefore focus on prison administration and political turmoil, completely neglecting local peasant life. Shame is the running theme in French Guiana's history. However, I would argue that the region should be proud of specific aspects of its history, such as the remarkable courage and willingness of the Creoles to live in the savannas and develop the coastal lowlands. With no assistance, the freed slaves established family farms on uninhabited land. Despite meager starting conditions, they succeeded in developing settlements and a strong sense of community, despite their isolation. The paucity of the historical records that address this segment of Guianese society requires that we turn to other sources. Archaeology, and to a lesser extent, ethnography, help to fill these historic lacunae.

Creole Settlement of the Lowland

Twenty years ago, I recorded the remnants of several Creole settlements in the savannas and located them on aerial photographs (Rostain 1991). Later work carried out in 2008 and 2009 then revealed the full extent of the Creole settlement patterns. A zone of approximately 2,000 hectares, west of Kourou, was carefully and systematically sampled with test pits. This zone was home to most of the Creole community prior to the establishment of the Centre Spatial Guyanais.

Comparison of archaeological, ethnographic, and cartographic data illuminates the earlier lifestyle in the region. Two roads structured the settlement; the first was the Anse Road that ran along the beach, and the second was the former route of the Route Nationale No. 1, which ran along a *chenier* further south. The farms, scattered on the sandy elevations between "the Savanna" and "the Anse" settlements, were connected to these two axes by crossroads.

Furthermore, a large creek and an artificial canal, parallel to the shore, structured the landscape and affected the distribution of Creole settlement. The Leroy Canal, built by slaves to drain the marshes, flows northwest from the Kourou River, parallel to the shore, crossing 11 kilometers of savanna. The creek runs south through a sandbar. All of the residential units are located at least 800 meters (often less) from one of the two channels. The northern end of the Bois Diable *chenier* ends in a narrow tongue 20 meters wide before being separated from the Colline Sable *chenier* by a stretch of marsh 100 meters wide. Two convex canals cut across this peninsula, connecting the two marshes that surround the *chenier*. They measure 3 meters wide and 0.5 meters deep, with a wide-mouthed U-shaped profile. The question arises as to whether these canals were created by indigenous peoples or Creoles. Given that they continue south as channels that flow into the Leroy Canal, they cannot predate its construction in the nineteenth century. Moreover, these types of earthworks on dry land are not typical of pre-Columbian sites. It is therefore logical to conclude that they date to the colonial period and are contemporaneous with the Leroy Canal. While it is certain that the construction of these canals required significant effort, it was justified. They served as a shortcut boat route from one marsh to the other without having to navigate around the Bois Diable isthmus.

Fourteen Creole domestic groups were discovered and studied over the course of two years of survey (Clerc and Déodat 2009). Although they are grouped together to form loose agglomerations, the units are still separated from one another by 200 to 600 meters. This was not the case throughout the coast because settlements become progressively more distantly spaced apart the farther one moves west of Cayenne. Thus, the closest neighbors of Raphaël Létard's (2007) parents' farm, close to Sinnamary, lived more than 3 kilometers away. Indeed, while specific events occasionally drew the community together, individualism was the rule rather than the exception in the Creole world, which was still traumatized by centuries of slavery (Jolivet 1982). Families were self-sufficient. This settlement, although dispersed, was made up of four entities between Kourou and Sinnamary: Anse, Savane, Karouabo, and Malmanoury, the latter two depending on Sinnamary.

Each farm spread over an area of 100 by 50 meters to 300 by 150 meters. The residential structure itself was simple, either constructed directly on the ground or on low stilts that rested on a brick or rock base. The *grignon franc* (*Sextonia rubra*, Lauraceae) was preferred for the exterior walls and *simarouba* (*Simarouba amara*, Simaroubaceae), lighter in color, for the flooring (Létard 2007). Test pits in Creole sites west of Kourou revealed alignments of bottles buried upside down. This practice of delimiting exterior spaces and decorating the flowerbeds that border the houses with the bases of glass bottles persists today among the Creoles of French Guiana. All of the households contained the same fittings, and only 2 of the 14 settlements possessed a *couac* (manioc flour) oven, installed close to the kitchen that was located a dozen meters from the house. The residents of these structures likely possessed elevated social status. The oven functioned on the same principal as flour mills in Europe; either the families came to cook their manioc flour, or the owners of the oven produced the *couac* to redistribute to their neighbors (Clerc and Déodat 2009).

The persons interviewed, now living in urban zones, were nostalgic about this form of mutual aid when recalling their lives on the savannas (*Je me Souviens* 2000). Animals slept in enclosed spaces: the henhouse (carefully protected against predators), pigsties, and stables. One or more wells were excavated around the house and the kitchen garden. Domestic wells could be dry outside the rainy season, so during the dry season the community often relied on collective wells that were located 1 to 3 kilometers away from the houses. These wells were sometimes paved with stones to retain the water. Domestic wells measured 1.2 to 1.5 meters in diameter by 2.5 to 3 meters deep. A surrounding wall was built to prevent children and animals from falling in (Clerc and Déodat 2009). The fields were distributed around the settlement in an elevated zone to prevent flooding.

According to the census of 1961, the community of Kourou was home to 659 inhabitants, of which the vast majority lived outside the village in 180 small farms spread out over the savannas, principally along the two coastal roads (Institut National de la Statistique et des Études Économiques [INSEE] 1962). According to accounts from the period, about 50 families were settled in the Savanna sector, close to Route Nationale No. 1, and between 15 and 20 families lived along the Anse Road. Given the official census, it is likely reasonable to double those numbers. The depopulated village served basically as an administrative center and a gathering place for community events (worship, trade, festivals, etc.) (*Je me Souviens* 2000).

This agricultural world disappeared suddenly in 1965 with the establishment of the Centre Spatial Guyanais. The resulting urban center exploded demographically with the inclusion of rural populations and city employees. "A village was forced to give birth to a city. A painful parturition. Scars which are still sensitive" (*Je me Souviens* 2000:4).[3] A single holdout, Miraka, refused to move, remaining in Anse until he died. The bluntness of the expropriation left the inhabitants with a bitterness and frustration that still lives today. "Unfortunately the base was established and since them we've known the misery of slavery" (*Je me Souviens* 2000:26).

Today, the territory of the Centre Spatial Guyanais is a closed zone, but the administration turns a blind eye to the hunters that enter in the woods, abundantly repopulated by a variety of fauna since the abandonment of the Creole settlements.

Creole Agriculture

It is generally believed that Creole farming is limited to shifting agriculture, inherited from the indigenous traditions. No mention of earthworks is found in the texts:

> In the proximity of the settlement itself, with its wood cabin, its shelter for the kitchen, its "gragiri" where the couac was made, its henhouse, all arranged around a large courtyard that played a central role in the organization of daily life, close to the settlement therefore is found the orchard, which generally contains 30 to 40 trees (citrus above all, and also coconut trees). . . . A bit farther away was the small slash-and-burn field of "vegetables" (bananas, *dachine*, often with a corner dedicated to vegetable crops) carefully enclosed while the manioc slash-and-burn fields were cultivated in the highlands, following the traditional technique of shifting slash-and-burn agriculture. . . . The average farm covered 2.5 to 3 ha, the practice of slash-and-burn agriculture nevertheless implying a much larger area (Jolivet 1982:448–9).

Except for the very complete work of Marie-José Jolivet (1982), there are very few texts that recount the life of Creole peasants in the coastal savannas. Nonetheless, there exists the charming work of an old Creole who describes his childhood in the early twentieth century around Sinnamary (Létard 2007). This lively narrative discusses daily life and important events related to a family that occupied a *bitasyon* (settlement) on the left bank of the Sinnamary Estuary. For example, the author describes the clearing of a new slash-and-burn field by the organization of a "*mayouri*" or communal project that gathered several neighbors together to carry out a difficult task: "The work was hard. The men perspired. They were happy to be together and to show one another their savoir-faire" (Létard 2007:25). He also discusses a supplementary farming technique that is unique and not previously reported: "The vegetable garden was made up of boxes full of earth that were supported over the lake by stilts. In these boxes were planted scallions, celery, parsley, and cabbage. This technique was the surest method to defeat the manioc ants and other insects destructive to vegetable crops" (Létard 2007:23–4).

Due to the lack of information on Creole agriculture, we must turn to other disciplines to reveal these forgotten practices. The first remains of Creole landscape modifications were discovered in 1989 during my ultralight aircraft flights and the study of IGN imagery. Stereoscopic analysis of aerial photos reveals the different human modifications to the environment. A good example of pre-Columbian and Creole remains can be seen in the snapshots IGN 003.100 Nos. 52 and 53 from 1955, just west of Kourou, north of the modern airfield (Figure 61). These high-quality photographs, at a 1:10,000 scale, were taken to create precise cartography for the future Centre Spatial Guyanais. The local farmers passed their last final years of freedom in the savannas before being expelled from the zone and relocated to the town of Kourou. These images therefore represent the last snapshots of a bygone era.

Careful analysis of the photographs highlights two cultural ensembles that are distinguished from one another not only by their locations, but also by the techniques that were used. There are both pre-Columbian modifications and evidence of activities and constructions carried out by Creole farmers.

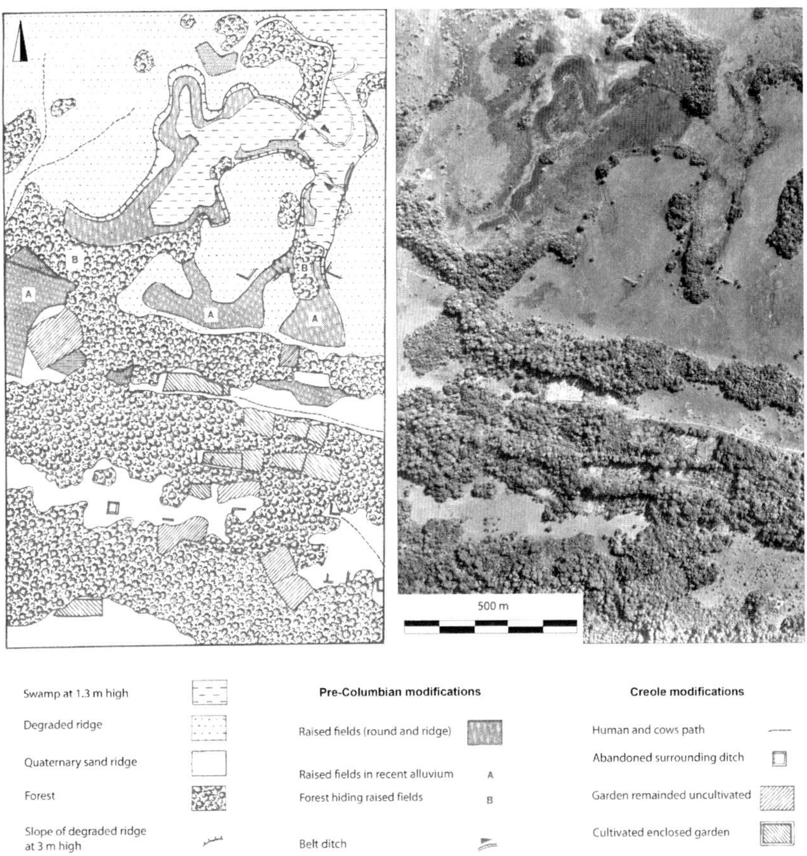

Figure 61. Stereoscopic interpretation of photos IGN-003.100 Nos. 52 and 53 at a 1:10,000 scale, west of Kourou just prior to the abandonment by the Creoles and the establishment of the Centre Spatial Guyanais (photo IGN 1955).

Pre-Columbian Modifications

Pre-Columbian raised fields are located between the coastal sandbars that average 3 meters high, and the matrix that is 1.3 meters deep. This placement ensures that the fields are in a zone that is neither overly flooded nor too dry. Elongated raised fields are found in the higher areas constructed parallel to the slope to retain water, whereas large, circular mounds are found in the marshes. To the north, raised fields appear to be organized in homogeneous sectors of a half-hectare each, covering a total of 5 hectares. Each of the small groups may perhaps correspond to a family's production. In the clayey thalwegs, the small raised fields have been buried by colluviation of the mounds themselves from the sandbars, or even the *cheniers*. A general leveling of the model is observed in some areas. Some raised fields are currently masked by the forest. After their abandonment, alluvial deposition allowed the forest to advance. This phenomenon of forest regrowth on the savanna in the absence of human activities is observed at a number of sites on the coast.

The enclosing ditches appear perpendicular to the path of the water's withdrawal. They serve to capture excess water in the marsh during the rainy season flooding, thus protecting the raised fields from submersion and pollution from outside elements.

Creole Modifications

Evidence of recent Creole settlements is not found on the higher elevations. A dozen slash-and-burn fields are uniformly located on the sandbars, where the soil is of better quality. Some, recently abandoned, have been colonized by abundant secondary vegetation. Other slash-and-burn fields are still used and fenced, likely to keep out livestock.

The visible trails on the nonwooded sandbars were produced by the cattle that occasionally come to graze the zones that are more than 4 meters high.

The quadrangular enclosing ditches were not in use during the period when the photography was carried out. They are located close to zones that flooded rapidly. The closed squares appear to have surrounded a house, while the more open configurations protected agricultural beds from excess water.

It is evident in the photography that the modern inhabitants do not cultivate the seasonally flooded savannas, and that they do not even cross them. During survey, numerous remains from recent habitations were recorded in the forest (masonry, bottles, ceramics, etc.). These domestic sites were nonetheless already abandoned by the Creoles when the photos were taken, and only a few slash-and-burn fields were still cultivated.

Archaeology significantly changes the image of a strict use of shifting agriculture. Since 1989, using aerial imagery, I discovered ditches that form right angles on the sandbars (Rostain 1991). This type of structure is clearly distinct from pre-Columbian landscape modifications not only by their shape but also their location. Pedestrian survey revealed a systematic association with late colonial remains: walls, glass bottles, ceramic, metal, etc. The most remarkable structures are the canals or ditches, sometimes quite large, located on sandbars and ridges. I sometimes call the basic configuration of modification the "Creole Square" (Rostain 1991). Ditches are 1 to 1.5 meters deep, with the berm on the inside, and encircling the settlement (Figure 62). They form a square 30 meters on each side that surrounds the house and some fruit trees: mango (*Mangifera indica*), soursop (*Annona muricata*), golden apple (*Spondias dulcis*), lemon (*Citrus limonum*), guava (*Psidium guajava*), etc. These ditches serve two main functions. The first is to keep the soil dry while also forming a reservoir. The second use is to protect the domestic space from the intrusion of livestock. Some ditches are interrupted by passages (formerly closed by a fence) allowing access to the house. Other types of ditches, always straight but otherwise of a variety of shapes, are found on the farms. Variable in length, they consist of a simple line, two lines at right angles, or a maze of channels. Like the Creole Squares, they were located on the higher zones out of the flood zone. Their function was not to protect a zone from livestock, but rather to improve drainage, particularly for the groups of small agricultural beds.

In the 1980s, none of the former inhabitants of the Kourou savannas mentioned these structures. They seem to have disappeared from collective memory. Nor is this type of ditch described in the archival texts or in historic studies. Only M. de Préfontaine alludes to them in the second half of the eighteenth century:

> Heavy rains can cause him to lose his harvest, if he does not take precautions . . .). To avoid this disadvantage, it is necessary to elevate the soil of the garden above the field that surrounds it, so that each square or flower-bed is a foot higher than the soil of the garden. . . . Independent of these precautions, a ditch two feet deep by at least four feet wide should be excavated around the garden, to allow the water to more easily drain away (Préfontaine 1763:18).

Aerial stereoscopic photography reveals the Creole ditches as well as other contemporaneous structures. These latter include small rectangular mounds arranged in a checkerboard pattern. Excavations at Sable Blanc and survey carried out around Kourou indicate they are Creole agricultural beds. They likely correspond to those referred to by Préfontaine as "squares or flower-beds." If this is the case, this farming technique existed for at least two centuries in French Guiana, and possibly in the neighboring Guianas as well.

An interdisciplinary field study has been recently carried out on the Creole agricultural beds. A small complex associated with the

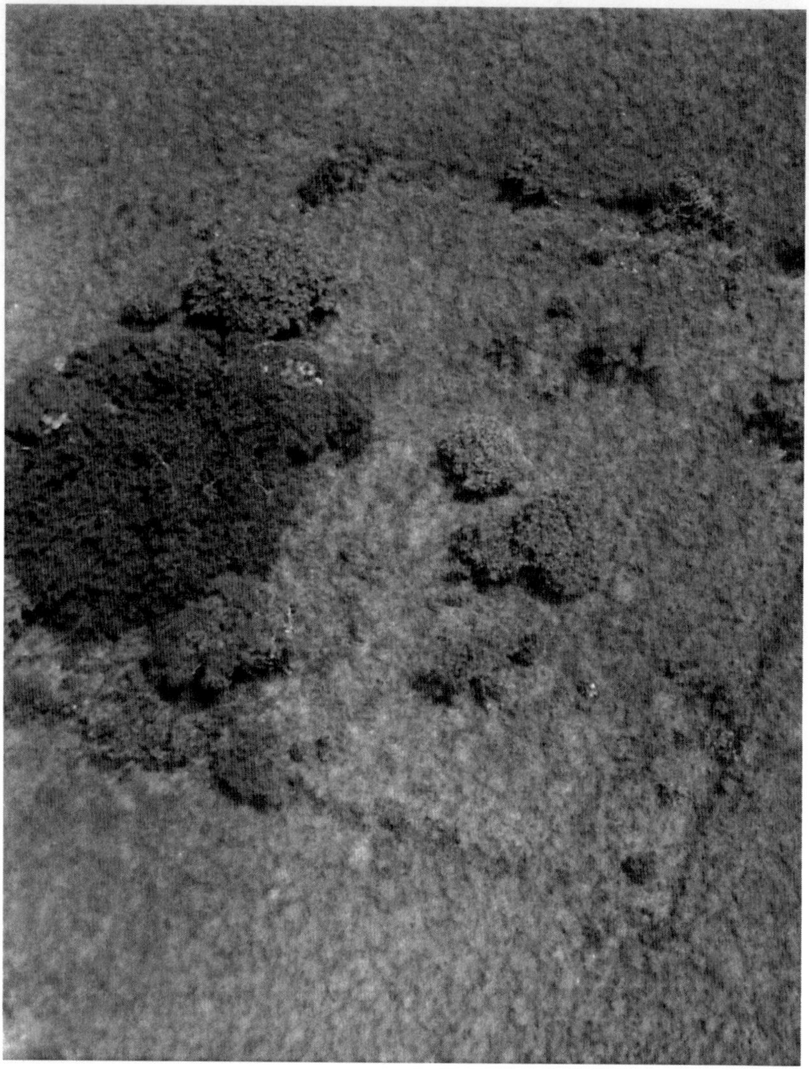

Figure 62. Ancient Creole enclosing ditch near Kourou, French Guiana (photo by Stéphen Rostain).

pre-Columbian site of Sable Blanc was intensively studied by archaeologists, pedologists, and archaeobotanists from the "Archaeology and Ecology of the Coastal Savannas of Guiana" consortium.

The mounds are located on the edge of the Barbakoeba site of Sable Blanc. They cover a small area covered by forest that lies between a road running perpendicular to Route Nationale No. 1 and a small embankment that encloses the site's pre-Columbian cemetery. They are difficult to recognize in the forest and are only noticeable through regular changes in the ground level that are detected when walking over the survey area. They became completely visible in a parcel that was cleared for planting in 2006. I undertook a study of these features during excavations at Sable Blanc in 2007. Laid out in a checkerboard pattern, they are small rectangular beds, occasionally square, from 0.5 to 3 meters long, 0.5 meter wide, and 0.3 meter high, separated by canals 0.5 meter wide (Figure 63).

The first step in this fieldwork involved coring for soil samples. Pedologists Bruno Glaser and Jago Birk from the University of Bayreuth (Germany) determined that none of the soil horizons detected showed hydromorphic characteristics. This location has therefore always experienced oxic conditions, which means that it was never frequently flooded and that the soil water table is generally deeper than 50 centimeters. We can thus deduce that these raised fields were never in a humid or seasonally flooded environment, which would explain why they are not very high. They were thus not constructed to resolve problems of drainage, but simply to concentrate the most fertile soil around the plants. In the 1980s, I detected some raised-field sites in other parts of French Guiana, also under forest cover in non–floodprone areas (Rostain 1991). The area was very likely once covered by trees that had to be cleared

Figure 63. Agricultural beds on the Kwatta-Tingiholo site, south of Paramaribo, Suriname (photo by Laddy van Putten).

to construct the raised fields. A fine level of small wood charcoal fragments ran along the base of the mounds. It seemed to correspond to a controlled burn of the forest, intended to open a clearing.

The raised structures occupied a small zone circumscribed between the embankment running along the pre-Columbian cemetery and the modern road. It was unclear whether these features are pre-Columbian in origins. The complex is much smaller than others on the Guianese coast. The length, width, and height of the fields themselves are also smaller than those of other coastal raised fields. Finally, slash-and-burn used in tandem with small agricultural beds does not correspond to a known pre-Columbian practice, but rather to the techniques of the modern peasant farmers of Guiana. The size and shape of these mounds recall those of the former Creole agricultural beds that are found on the savannas, rather than those of the pre-Columbian mounds. It is thus more likely that local farmers built them recently. Dating of several samples of charcoal collected from the fire level below the beds confirmed this hypothesis. All of the dates returned were modern, corresponding to the twentieth century.

Archaeobotanists José Iriarte and Jennifer Watling selected four beds for sampling for phytolith analysis. They determined that maize had been cultivated on these beds (Iriarte et al. 2010). Although relatively infrequent, maize is one of the plants grown today by Guianese peasant farmers.

The ancient agricultural beds discovered in the savannas are always small. They rarely measure more than 4 meters long and 1 meter wide. The technique has not died out; it is still practiced on several family farms, particularly in Suriname (Figure 64). There, the beds vary significantly in size. They may be very long, parallel beds several meters long as on the fertile sand and shell ridges of Kwatta, south of Paramaribo. Others, for example among the Saramaka in eastern Suriname, are rectangles a few meters on each side constructed in sandy zones. It is interesting to note that the abandoned modern beds are well conserved, particularly in the forest, such as those of the *cheniers* of Kourou, unused for more than 50 years. In contrast, those arranged on the open savannas tend to suffer significant degradation in a short period of time.

Agriculture in the Savannas Today

The economic policies of Guiana vary considerably from one country to another. Guyana and Suriname focus primarily on bauxite mining and farming of polders, whereas French Guiana, with the financial support of the city, intensively developed its service sector and relies on spillover from the activities of the Centre Spatial Guyanais. For example, barely more than 23,000 hectares are cultivated in the entire country.

Figure 64. Recently abandoned Creole agricultural beds near the pre-Columbian Sable Blanc site, French Guiana (photo by Stéphen Rostain).

Half of the farms do not exceed 2 hectares, in which vegetables, tubers, and fruits are grown through slash-and-burn agriculture, principally for direct consumption by the family. Close to 40 percent of farms are of medium dimensions, between 2 and 10 hectares. They are located on the coast, close to settlement clusters. Large farms (more than 10 hectares) occupy approximately 60 percent of the country's overall cultivated area though only representing 3 percent of all farms. Located on the coastal plain, they comprise orchards, savanna pastures for cattle, or rice paddies in polders.

The Use of Savannas

At the end of the 1980s, I interviewed several residents of Kourou who had previously lived in the surrounding savannas, as well as elderly persons who knew the coast before the 1950s, to obtain information regarding life in this environment and its use by the inhabitants. Several additional projects with similar objectives were subsequently carried out during the "Archaeology and Ecology of the Coastal Savannas of Guiana" research (Clerc and Déodat 2009; Schnier 2009). The ethnic groups consulted include Creoles, Haitians, Kali'na, Saramaka, and city-dwellers who have lived for a long period of time in French Guiana. A central goal of these investigations was to obtain information regarding the knowledge of raised fields in the marshes. The responses were varied, but most interviewees had no knowledge of the existence of the mounds, or had forgotten them. Twenty years ago, my informants generally believed these structures to be natural in origin and often told me: "The Lord made them." When the Piliwa complex was discovered at the beginning of this millennium in the community of Awala, it was difficult to convince the Kali'na that they were raised fields; they considered the mounds to be a natural formation. The indigenous groups of Kourou hold the same opinion: "No, they are natural everywhere. . . . Nature often makes things that we do not understand. . . . From father to son we would know if the mounds had been used" (reported by Schnier 2009:24). In all cases, the interviewees did not believe that the mounds could be the works of pre-Columbian populations, who were considered to be nomadic hunter-gatherers. They therefore could not have built on and modified these immense areas. The most extreme remark came from an elderly Creole from Iracoubo, who affirmed that before the arrival of the "Guianese" (Creole people), the indigenous people did not know *couac* (the flour made from bitter manioc), and that it was the "Guianese" who had taught the indigenous population its production. The direct accounts from local residents must be interpreted with caution because they can be marked by cultural assumptions and prejudices. The objective is therefore not to collect the truth, but rather to understand the local perceptions of specific phenomena and of the landscape in which the population developed.

One of the most widely held convictions among the inhabitants is that the savannas are sterile, although this belief is not based on empirical experience. They cross these areas, but do not settle or even farm them. "The savanna gives nothing. . . . Just to make the grass grow, we must add fertilizer. . . . People cannot cultivate on such poor land" (reported by Schnier [2009:25, 29–30]). The rejection of these seasonally flooded areas is particularly visible in aerial photography.

Today, farmers in Guiana sometimes build small, elongated raised fields and mounds, much smaller than those of the ancient indigenous groups. They are not necessarily located in wetland zones, and their principal function is to aerate the soil and concentrate nutrients. In the Lower Mana, toward Acarouany, the Saramaka gather the surface earth in small mounds around their plants on a surface not subject to flooding. Creole and Haitian farmers also practice this method in the seasonally flooded zones at the foot of the Montagne du Mahury, on Île de Cayenne.

In the savannas, some farmers occasionally plant on small raised fields. On the border between the Organabo River and savanna, Muriel, a Saramaka native of Suriname, cultivates on mounds using slash-and-burn technique. She rotates between crops of sweet potato and bananas, and fruit trees occupy the lower area (Schnier 2009). Moreover, in the savanna, 10 years ago she used a tractor to build small, round raised fields next to large pre-Columbian polygonal mounds. The sweet potatoes she tried to cultivate did not produce the expected returns, so she abandoned them after three months to concentrate on the slash-and-burn fields.

Another attempt to farm close to 2 hectares of raised fields was carried out a few kilometers east of Organabo. From 2000 to 2004, an independent farmer, Kévin, achieved three harvests of pineapples before being evicted by the municipality for occupation without a title (Schnier 2009).

Farmers who had arrived from France attempted a long time ago to cultivate small raised fields in the savannas. During our 2003 survey, we recorded several elongated raised fields just south of the Kali'na village of Bellevue, 6 kilometers west of Iracoubo. They were arranged perpendicular to the slope of a sandbar in a non–seasonally flooded zone. In a more recent ethnographic investigation, a French farmer indicated he had built these features in 1996 and cultivated them with the indigenous people for a year (Schnier 2009). During the same surveys, elongated raised fields had been discovered in the Corossony savanna, to the south of Route Nationale No. 1. They were also distributed along the sandbar, but oriented perpendicularly to the slope. A resident of the area said that he constructed them in the early 1990s after having read my publication about raised fields (Rostain 1991). Nevertheless, he had not considered them adequately sustainable to justify continuing his experiment.

Finally, during flyovers, I recorded several fairly extensive areas of recent modifications, generally with the aid of construction machines. These are generally drained parcels similar to small polders.

Haitian Raised-field Agriculture

Recent Haitian immigrants in French Guiana have mastered the technique of raised-field agriculture in nonflooded zones (Figure 65). Nonetheless, these groups do not possess a pre-Columbian heritage. Rather, they derive their farming skills from the cultural memory of agricultural techniques practiced in Africa (Pélissier 1966). These techniques are commonly used in Haiti; the zone surrounding Port-au-Prince is covered with mounds.

In Haiti, peasant farmers have developed a system of subsistence farming centered around their homes. They distinguish their fields according to three types of "gardens" with relation to their distance from the house (GRET 1990): a *devant-porte-kaye* (front door) garden from 500 to 2,000 square meters' a *pré kaye* (close) garden from 1,000 to 5,000 square meters, and one or more *loin kaye* (distant) gardens.

The *devant-porte-kaye* garden, which directly encircles the home, is reserved for fruit trees, vegetables, and herbs. The *pré kaye* and *loin kaye* gardens, found in nonwooded or sparsely wooded contexts more distant from the home, are used for growing beans, maize, yams, sweet potatoes, and manioc. Raised fields are constructed in the latter two types of gardens (GRET 1990). After being hoed, the weeds are separated from the earth and set out to dry for two weeks. They are then gathered together and buried with the earth excavated from the top

Figure 65. Modern Haitian raised fields near Kourou, French Guiana (photo by Stéphen Rostain).

15 centimeters of the ground surface to construct the mounds. These structures measure on average 1 meter in diameter and 80 centimeters high. The buried weeds are concentrated at their base, covering a quarter of the mound's circumference. This method improves the aeration and drainage while concentrating the organic material. Although the mounds are arranged side by side, the different plant species are not planted uniformly across them, but rather are distributed in a heterogeneous pattern according to spatial variability in the level of fertility. Maize, the most demanding plant, is planted above the quarter where the weeds are buried. Yams and sweet potatoes are both planted at the top of a mound in an alternating distribution (i.e., mounds with yams alternate with mounds with sweet potatoes). "Congo pea" (*Cajanus cajan*) occupies the canals between the mounds. Beans are planted on the mounds and in the canals. Finally, manioc is planted on every second mound.

The seasonally flooded northern coastal plain is also farmed, particularly for rice, but with a substantially greater workforce, exceeding the scale of the family farm. In this environment, similar to the coastal lowlands of Guiana, Haitians practice agricultural techniques that resemble those of the Guianese Arauquinoids. In the dry season, when the plain is not flooded by the rising water table, they construct large mounds and elongated raised fields. Root crops thrive in these well-aerated soils, while rice flourishes in the canals that run between the mounds (GRET 1990).

The technique of *pré kaye* and *loin kaye* mounds was imported to French Guiana. In 2005, I noticed several raised fields being cultivated on the western outskirts of Kourou along the Bois Diable sandbar. Intrigued, I searched for the farmers and located a Haitian woman who cultivated these fields. She began to construct these mounds in 2001, planting bananas, manioc, sweet potatoes, and beans. She had clearly noticed the pre-Columbian raised fields close to the marshes, which she attributed to her predecessors. During the following years, I observed several other farmers who arrived in this zone and established small operations using similar agricultural methods.

During my excavations at Bois Diable, I regularly interviewed Mr. Lucian, a Haitian who had recently come to live on the western edge of this mound complex. He used the terms "*buttes*," "*hautes terres*," and "*gonfle terres*" (mounds, high lands, swollen lands) to describe these fields to Matthias Schnier (2009:35), who also spoke with him. The shapes of the mounds are varied: round, elongated, straight, or sinuous. The round fields measure from 50 to 100 centimeters in diameter and between 20 and 60 centimeters high (Figure 66). Their two principal crops are bananas and manioc, but beans, sweet potatoes, yams, and Congo peas, etc. are also grown.

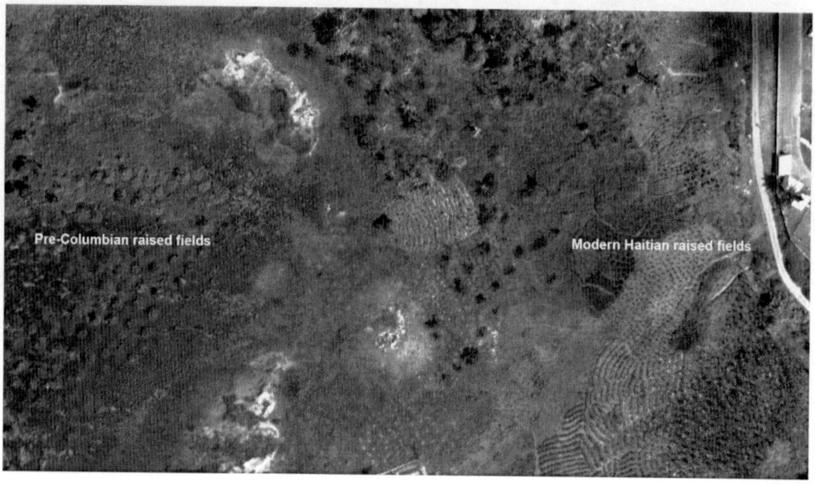

Figure 66. Modern Haitian raised fields close to pre-Columbian raised fields, near Kourou, French Guiana (photo by L'Avion Jaune).

Despite clear differences existing between Haitian and pre-Columbian raised fields, ethnographic study of the modern farmers is very enlightening. For example, I once observed Mr. Lucian digging a twisting canal in the hard, compact earth at the center of his mounds. I asked why he was undertaking this extra (and in my eyes, unnecessary) effort of this excavation that seemed to run every which way. It appeared that it would be much simpler and faster to dig a straight channel. Mr. Lucian patiently explained that a straight ditch would be much less useful because the zigzagging course of his canal between the mounds would irrigate more surface area and, more importantly, it would slow the flow of the water that would otherwise move too quickly to be absorbed by the soil. This is probably also the reason that many pre-Columbian canals follow a sinuous line (Figure 53).

Disappearance and Preservation of Raised Fields

Phenomena of Erosion and Destruction

Two principal and natural reasons explain the modern phenomena of the complete disappearance of some structures or their colonization by the forest: first, a process of alluviation and colluviation, followed by an absence of maintenance. Another cultural reason is the direct destruction by modern activities like agriculture, building, earthworks, country planning, etc. (Figure 67).

The first cause is modifications of certain parts of the ancient coastal plain since the period of raised-field construction. In the sector north of

Figure 67. Obliterated raised fields by modern earthworks in the Renner savanna, near Sinnamary, French Guiana (photo by Stéphen Rostain).

the Montagne des Pères near Kourou, important differences are evident between the aerial photographs taken in 1955 and 1987. In 1955, several alignments of mounds as well as enclosing ditches and an earthen levee appear clearly in the marsh at the edge of the sandbars. In 1987, the same mounds are almost invisible, whereas the enclosing ditches and the levee have completely disappeared due to either sedimentation or erosion. A flyover of this area in an ultralight aircraft in 1989 confirmed the destruction of these features.

Since raised fields were well preserved for several centuries, the sudden disappearance of some of these structures the last 30 years might at first seem astonishing. The explanation is found in many recent public building projects. Several years ago, construction was undertaken to deviate Route Nationale No. 1 south of its original trajectory to link it with the new national highway. A north-south road was constructed from mounded earth that blocked off the marsh transversally between the *cheniers*. This road would have influenced considerably the hydrological network of the area, creating new checks on the flow of water and sediment, perhaps alluviations or colluviations, which would have progressively led to the destruction of the mounds. A similar phenomenon is observed on the outskirts of Kourou, where a similar deviation of the former national highway was carried out.

On the western part of the recent coastal plain of Suriname, the construction of the east-west road, around 1964, modified the local hydrology. Due to the absence of *cheniers* in this region, a small dike was built to support the road. After a few years, the drainage canals became blocked and the water could no longer circulate from north to south. No longer affected by the tides, the southern marsh became an exclusively freshwater environment. To the north, where the freshwater could no longer flow, conditions became increasingly more saline. In addition, the lack of water flow led to the accumulation of sediments, which progressively began to bury several groups of raised fields. Comparisons of aerial photography between 1947 and the present day clearly demonstrate this phenomenon of destruction.

The second explanation for the current sedimentation of some raised fields in the highly seasonally flooded areas is related to the circumstances of their abandonment. Although erosion does not generally play a significant role in modifying the early coastal plain of Guiana, precipitation is nonetheless a principal factor in changes to the ground surface. The surface of some raised fields is irregular and broken into clods. The most impressive example is found in Savane Maillard, south of Macouria, where the mounds themselves are divided in two to four (or even more) micromounds. They were originally alignments of square mounds averaging 280 centimeters per side and separated by canals that average 70 centimeters wide.

This erosion is increased by the annual fires on the savanna that destroy the plant cover, which normally protects against soil loss. A pedological test pit at the foot of a mound illustrates the general process of leveling raised fields. The earth of the mound tends to erode into the intermediary canals, resulting in erasure of the reliefs and canals. Regular maintenance of the raised fields is necessary to clean out the canals and replace the organic sediments on top of the mounds. A gradual leveling of the mounds began with the cessation of agriculture, perhaps accelerated by more recent anthropogenic processes (burning of the savannas, constructions).

Preservation of the Structures

Although the erosion and disappearance of the raised fields are not surprising, their previous preservation for centuries appears extraordinary. For this reason, the excellent preservation of pre-Columbian mounds across the Guianese coast is astonishing to many. One explanation is that human impact has been relatively minimal on the savannas since their abandonment (although this may change if the attempts at rice agriculture on the savannas, such as those close to Iracoubo a few years ago, become more common). Only the fires occasionally set by the modern inhabitants of the savannas cause damage (on savanna burning, see also Iriarte et al. [2012]). By removing the vegetation, they leave the mounds unprotected against the erosive action of strong rains (Rostain 1991). For example, the inhabitants of Savane Maillard, east of Macouria, periodically burn the grass during the dry season. As a result, although the raised fields are still visible from the air, they are in large part destroyed on the ground and more particularly in the higher areas, which are less flooded during the rainy season.

Standing water has little effect on the mounds, except when dikes are constructed to accommodate a road. In this case, it is possible to see colluvial deposition that can obliterate the mounds. In contrast, structures that are flooded for most of the year are often the best preserved.

Nevertheless, these different elements do not explain the preservation of the mounds for almost a millennium. Indeed, the 2 to 4 meters of annual rainfall typical of the Guianese coast seemingly should have totally eroded and destroyed the raised fields. Despite the collaboration of earth sciences, such as pedology, geomorphology, and hydrology, archaeologists remained puzzled by this enigma. Some theories were advanced (the role of plant cover, standing water, etc.) but no satisfactory solution had been found. It took the cooperation of ecologists to reveal the incredible scale of the contribution of natural agents to the maintenance of raised fields.

Ecologists of the "Archaeology and Ecology of the Coastal Savannas of Guiana" project explored the hypothesis that the actions of "natural ecosystem engineers," such as ants, termites, and earthworms, help to counter erosion. Their study focused on the impact of ancient human modifications to the landscape on the modern dynamics of the savannas. They relied on descriptors such as the biological functioning of soils, the ecology of wild manioc, the ecology of ants, and the interactions among these different organisms. Indeed, these organisms prefer to colonize the ancient mounds, the only patches in these landscapes where the soils are relatively well drained. The ecologists observed that the activity of soil-dwelling social insects and earthworms were extremely important in the raised fields, each mound housing at least one ant-nest because, as in Bolivia (Denevan 2001), these insects prefer the well-drained soils of these elevated zones for their settlements. A similar phenomenon can be observed in Argentina, where *Camponotus punctulatus* ants build *tacurúes*, very hard anthills that average 40 centimeters high (Folgarait et al. 2003). Nevertheless, these ants prefer to settle aerated soils and thus often invade abandoned rice paddies. They invade these largely because these soils, formerly fertilized, produce lots of vegetation, giving them lots of insect prey (Doyle McKey, personal communication, 2012). Although the normal density of their nests is less than 40 per hectare, they can reach up to 2,000 nests per hectare in the former rice fields, where each anthill can reach 1.65 meters high and 2 meters in diameter in 15 to 20 years (Folgarait et al. 2003). In Guiana, during the high-water season, the ants seek refuge in the upper chambers.

The ants regularly carry subsoil from nest excavation to the top of the mound. Material carried to mound surfaces could replace material lost through erosion. *Acromyrmex* fungus-cultivating ants play a particularly central role in this mound-maintaining activity, moving not only subsoil but also transporting large amounts of organic matter to their nests, all on mounds (McKey et al. 2010; Renard 2007). Test zones have been observed and analyzed by ecologists, pedologists, and archaeobotanists for the last three years in the Grand Macoua and Organabo savannas. The data collected may eventually allow demonstration of the effective action of the natural ecosystem engineers and elaboration of precise models.

Furthermore, certain ants and termites transport organic and mineral materials to their nests in the mounds. These actions, coupled with the roots of plants (denser on the mounds than in the flooded matrix) modify the physical properties of the mounds' soil, rendering them more porous and resistant to erosion. Enriched by the accumulation and transformation of the organic material, the mounds are preferentially colonized by plants, which in turn deposit their litter on the mounds' surface. Similarly, earthworms are attracted by these aerated soils and the concentration of organic material. The activity of these engineers also modifies the properties of the soil. The increased porosity provided by the earthworms and other organisms increases the capacity for water infiltration, which is nine times higher on the mounds than in the surrounding matrix (McKey et al. 2010). Water infiltrates the mounds rather than running off of them so that erosion is diminished. By transporting material to the mounds and reducing their erosion, the communities of ecosystem engineers maintain the concentration of resources (and of organisms) on the mounds, and their depletion in the matrix.

These ecosystems thus seem to behave like self-organized natural ecosystems (McKey et al. 2010). This immediately raises the question of whether these natural engineers could have actually produced the mounds themselves. Ecologists have noted the absence of anthills on a zone leveled with a bulldozer around 30 years ago. They thus concluded that the ants and other engineering organisms were incapable of producing structures sufficiently elevated to escape annual erosion, particularly considering the short time interval between successive floods. The initial and rapid construction of the mounds by humans seems to be a necessary precondition for their colonization and maintenance by ecosystem engineers. However, in some seasonally flooded savannas elsewhere in South America, morphologically similar mound-field landscapes do appear to have been produced by natural engineers (Renard et al. 2011, 2012).

All of the processes undertaken by natural ecosystem engineers were examined in detail for three years by Delphine Renard, who presented a preliminary study as a master's thesis (Renard 2007) and a more detailed study in her doctoral dissertation in functional ecology at the University of Montpellier (Renard 2010).

Fungus-cultivating ants are not the only engineers that may play a role in maintaining the mounds. They are even absent in some complexes, particularly in the extremely flooded environments such as those west of Kourou. The first time that I tested a mound in the marsh close to Kourou, I found myself almost immediately covered with unidentified red ants. Since then, I have had several similar encounters working in the large raised fields found in zones that are flooded almost year-round. The sedges that cover the marshes are also invaded by these insects, which have rather painful stings. Red ant nests are often found in the dry islets that form the raised fields.

In contrast, the red ants, which in these marshy habitats are limited to the mounds, do not appear to function as ecosystem engineers like their fungus-cultivating cousins. They only excavate narrow tunnels, which do not emit much fill. It is therefore necessary to find other factors to explain the maintenance of these structures. Earthworms may have played an essential role. Similarly, plants constituted an equally important preserving element; the matrix is full of sedges, while grasses and shrubs dominated the mounds, acting as a protective layer against damaging climatic conditions.

To a certain extent, ants and other natural ecosystem engineers took over the role of maintaining the mounds from the indigenous people who initially constructed them, and have continued this function into the modern day. The combination of studies of archaeology, earth sciences, and ecology have begun to provide the multiple elements necessary to rigorously test the hypothesis that these landscapes were "co-constructed" by human and natural engineers.

One could argue that social mammals (humans) built the mounds, and then were replaced by social insects that maintained them. Thus, the mounds observed by the archaeologist proceed from a genesis that is both cultural and natural.

Conclusion

"East of Eden"

Before archaeologists began flyovers of Guiana Island, no one suspected that its indigenous people were capable of developing intensive agriculture. It was previously accepted that the "noble savage" only practiced extensive slash-and-burn farming (Figure 68). ". . . It consisted above all of the practice of abattis, a parcel of land in the forest that is cleared by fire and cultivated for a few years with food crops . . . since it is the only form of agriculture known by indigenous people, since ancient times" (Giacottino 1984:117). These "ancient times," fortunately, are becoming better understood, allowing us to discard prejudiced and a priori theories for a more informed and nuanced understanding.

The opinion of the new arrivals from the Old World regarding Guiana was often disparaging because it was tainted by eurocentrism and the repeated failures of the colonization:

> The largest reptiles are found here, and all the domestic animals here are of the puniest type. Livestock here degenerates; its milk is worth nothing, it lies down always in the open, in its own refuse, in the closed corrals; in winter, it has water and mud up to its chest. It must be enclosed, for fear of the tiger, and left in the open so that it is not exhausted by bats. They are so common and large in certain canton of Oyac and in the plains of Kau, for example, that they cannot be defended against. They attack the back, they cause it ulcers; the flies feed on the sores, they lay their eggs; worms arise; because here, all of the sores that remain exposed, are full of worms within twenty four hours; one can almost say that the plague never ceases in this country. The fish is rotten coming out of the water, the bread sprouts mold as it cools, the meat almost rots as it still twitches. The sky and the earth declare war on man, and he insists on settling and remaining here (Pitou 1805:311–12).

Many saw in Guiana (and the Amazon in general) an earthly manifestation of hell. The worst penalty for a slave from the West Indies was to be sent to Suriname (Stedman [1796] 1992), where the punishments were cruel and a slave cost less than a mule. Similarly, French Guiana was the land of the great punishment for convicts (Pierre 1982). This hellish vision of Guiana was also fed from the myths and practices of the indigenous groups. In the central savannas of Guiana, we still find today

Figure 68. "*Un abatis chez les Roucouyennes*" (a slash-and-burn field among the Roucouyennes [Wayana]) in the interior of French Guiana (Coudreau 1891:167).

one or two victims annually of the highly ritualized cannibalism practiced by Kanaimá shamans (Whitehead 2002, 2003). Humans become the prey of revenge killings and of supernatural beings whose demonic dimension is highlighted; the fear of these "devils" is still powerful.

On the other hand, others found the largest tropical rainforest in the world to represent a paradise lost, an Eden (Figure 69). In 1498, Christopher Columbus (Colomb 1992), upon reaching the Orinoco Delta, thought he had discovered a place "that could only be Paradise." Many others subsequently perpetuated this Edenic vision of the Amazon. Were they mistaken? The word "Eden," which gave "garden" to English, "*jardin*" to French, and "*jardín*" to Spanish, may derive from Sumerian "*e.din*," which means a cultivated land (Lemaire 1981). Considering the importance of pre-Columbian agricultural works that has been revealed today, the first conquerors of the New World were then not completely incorrect in seeing the Garden of Eden. Guiana lies east of this of this Amazonian Eden. But the region, surrounded by the Orinoco and the Amazon Rivers, which we call "Guiana," "Guiana Island," "Guiana Plateau," "Guiana Shield," "Eastern Amazon," or "Coastal Amazon," shows characteristics, both natural and cultural, that distinguish it from the rest of the Amazon.

Since the Conquest, Guiana continues to be perceived and colonized from two contradictory perspectives: a segregationist model derived

Figure 69. "Eldorado" is a good illustration of the recurrent myths of Guiana (Bouyer 1867:45).

from Protestant colonization versus an integrationist Catholic model (Lézy 2000). Between heaven and hell, between Catholic vision and Protestant consideration, Guiana was often approached from a religious perspective, ignoring its status as a geographic and cultural entity. Similarly, rationality often competed with myths in the accounts of those who attempted to understand Guiana (Hurault 2000). This marked dichotomy is found in the imagery illustrating Guiana since colonization. The urban world of cities built and maintained by Europeans is systematically contrasted with an untamed forest that is populated by "primitives" in shocking dress, always portrayed at their worst (Alès and Pouyllau 1992; Jolivet 2002; Lézy 1989). These perceptions have significantly shaped modern assumptions about Guiana.

Fairly quickly, some began to doubt Guiana's assumed fertility, which was based on the seeming luxuriance of the landscape. In 1823, Governor Pierre Bernard Milius questioned the reliability of positive assessments of the fertility of the tropical environment: "The first explorers, seduced by the vigorous nature of the forest, allowed themselves to be taken in by the misleading appearance of the vegetation. They did not realize that the richness of the vegetation was due not to the soil, which was composed almost completely of sand and pure clay, but rather to the ancient decomposition of a multitude of plants whose debris produced an excellent fertilizer." (cited by Turenne 2008). He was thus an ardent defender of cultivating the coastal plains, particularly through the use of polders, but he was halted by the magnitude of the task. Indeed, this "green Paradise" required a significant effort to become sustainable. Indigenous peoples understood this characteristic of their surrounding and thus worked to transform the coastal landscape over the course of several centuries. Many, such as Walter Raleigh, went mad chasing a myth. What became of the "El Dorado. Agricultural French Guiana" that Adam extolled in 1936?

Archaeologists possess few material clues to reconstruct pre-Columbian agriculture because these ancient practices generally leave behind few traces. The tools used to fell trees, clear weeds, construct mounds, or sow seeds are rarely well preserved because they are in great part made from perishable materials, particularly wood. The handles generally disappear, only leaving behind the active part of the instrument in the best of cases. Fortunately, analysis of microwear on the stone blades can reveal their original use (Rostain 1994a). Nonetheless, a few complete tools have been discovered that were preserved by anaerobic environments. Thus, for several years now, miners carrying out prospecting in French Guiana have recovered stone axes from river beds with their handles still attached. The most common of the tools, used for more than 5,000 years (Figure 70), are polished stone axes, which occur in a variety of forms, but with only two forms of hafting (Rostain 1990; Rostain and Wack 1987). Intense exchange for the stone used for the blades occurred between the source zones and regions lacking in the appropriate raw material. These networks

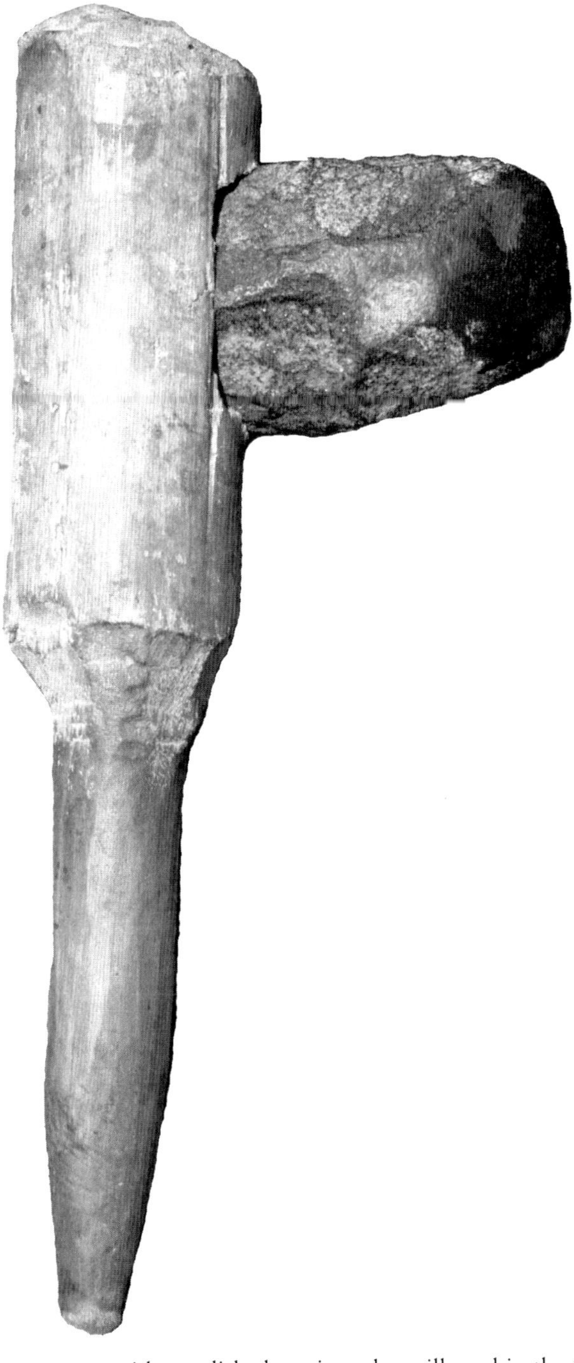

Figure 70. Stone axe with a polished cutting edge still used in the 1960s by an Akuliyo group in Suriname (photo by Stéphen Rostain).

were vital for many populations, who needed such tools not only to clear land for farming, but also for building houses, boats, tools, and weapons. The other types of tools are rarely reported because their ordinary appearance did not draw the attention of those who discovered them. The wooden shovel from the Prins Bernhard Polder site (Figure 56), or the turtle shell hoe from Hertenrits (Versteeg 1985) are rare examples. It is not difficult to imagine that digging sticks and wooden weeding blades were also used.

The study of indigenous agricultural practices, both in the highland and the lowlands, demonstrates that the adaptation to the environment and its possible transformation are more important factors for success than the simple productivity of the land farmed (Descola 1993). Therefore, the quality of the technological response is the most critical factor in the effective cultivation of almost any landscape. The virtuosity of the agricultural systems that have been developed around the world in extreme environments displays the great variety of technical solutions possible (Mollard and Walter 2008). Backed by more than 10,000 years of experience in the Amazonian rainforest, the indigenous populations developed multiple techniques of cultivation, sometimes combined, that were precisely adapted to the biotope exploited. There existed in the Amazon, and to a certain extent there continues to exist, a correlation between the great diversity of the environment and the multiplicity of the agricultural techniques. The pre-Columbian populations created a sustainable agricultural landscape. Charles Mann describes it as such: " 'Landscape,' in this case is meant exactly—Amazonian Indians literally created the ground beneath their feet" (Mann 2005:306). Although it is clear that multiple natural factors can explain the considerable botanical diversity of the Amazon, it is reasonable to propose a correlation between the more than 200 cultivated plant species in South America (Clément 1999a, 1999b; Denevan 2001) and the variety of agricultural systems developed by the pre-Columbian populations. Therefore, humans deliberately or unintentionally modified the natural composition of this tropical environment.

Although the human past in the Amazon at first glance seems to offer only a faint material record, when examined from the interdisciplinary perspective of historical ecology, a wide range of human impacts on the environment is revealed. It is often in wetland environments that these impacts are most evident. Indeed, one might assume that water would have an erosive and destructive effect. The reality is in fact the contrary; water stabilizes and preserves ecofacts. It is true that this environment has been more profoundly modified by indigenous peoples than has the interfluvial rainforest zone. Choosing to settle flooded or seasonally flooded regions required the populations to build earthworks to make these biotopes viable. It was necessary to balance the unique dynamic between earth and water. Thus, whereas irrigation was sometimes the goal (such as in the Andes), raised fields in the lowlands were often developed to deal with problems of drainage. They are generally located in seasonally flooded savannas or on the banks of rivers that were cleared for agriculture.

While shifting cultivation dominates today in the Amazon, it is now known that more permanent and intensive agricultural techniques were used during the pre-Columbian period. It is likely that few have left major remains behind. Yet, the simple presence of humans has left its impact on the environment; ancient human modifications to the landscape vary in their significance and visibility.

The range of these modifications is still greatly underestimated by the supporters of western progress. "These [indigenous] populations maintained a symbiotic relationship with their environment because their use never exceeded the ecosystems' natural capacity for regeneration; hunting, fishing, and gathering thus constituted their principal subsistence. It was the arrival of the first colonists that perturbed the system" (Transler et al. 2006:70). The impact of indigenous communities is often totally ignored, and only the effects of the western communities are considered (whether negative or positive). However, it has been demonstrated that the indigenous influence on the landscape began several thousand years ago and intensified shortly after the start of the Common Era. The areas characterized by pre-Columbian earthworks on the coast of Guiana are overall larger than those settled by the colonists since the seventeenth century, when the polders were constructed. The perturbations to the ecosystem by the construction of raised fields, canals, reservoirs, dikes, diversion of rivers, causeways, and settlement mounds permanently altered the environment. After the construction, cultivation, and abandonment of the raised fields, the Guianese savannas will never return to their original dynamic; the intervention of humans caused a perennial heterogeneity that has been maintained by natural ecosystem engineers (McKey et al. 2010).

The investment of these intensive transformations of the pre-Columbian landscape for the human populations was justified by the long-term benefits of these innovations. "*Le prix s'oublie, mais la qualité reste*" [Price is forgotten, but quality stays]. The investment required to build thousands of raised fields, causeways, habitation mounds, canals, and reservoirs allowed indigenous peoples to adapt the landscape to their specific needs for several centuries. In general, indigenous populations have tended to alter their environment, intentionally or not. The major modifications are the earthworks, both aggregations (e.g., mounds) and excavation (e.g., canals). Up until now, scientists have focused on the fertile flood plains of the Amazon and Orinoco Basins. Less attention has been paid to other prolific settings. The *várzea* was not the only fertile soil that could have given rise to stratified societies of the Amazon; the Atlantic coast and the wide Venezuelan and Bolivian savannas (*llanos*), as well as the volcanic piedmont of the Andes, were equally productive. A much higher density of complex societies developed in these areas than in the forested interior, and they constructed significant earthworks that are still visible today (Rostain 2008b, 2010a, 2012). Yet, we cannot simultaneously accept the existence of a mosaic of landscapes in the Amazon and reject the possibility of a diverse range of cultures coexisting. We must consider that the subtle variations from one group to another, still noticeable today, were very likely even more marked before European contact.

Moreover, researchers cannot realistically expect to encounter all of the ecofacts left behind because many have been degraded or destroyed through time. Many raised fields in the Maillard Savanna, French Guiana, like in Guayas, Ecuador, have been completely destroyed. Even the alignments themselves have sometimes disappeared because the mounds have been reduced to irregular, shapeless lumps that are no longer recognizable. Similarly, many canals have been partially or completely filled in with colluvial sediment. The study of pre-Columbian earthworks is particularly difficult in the Amazon, where more recent structures frequently have been built over more ancient ones. Furthermore, natural formations can easily be mistaken for human constructions, which were often originally influenced by observations of the natural landscape. Without archaeological dating, which is sometimes impossible to carry out, it is often difficult to verify the antiquity of features.

As has been discussed previously, human-constructed canals are particularly difficult features to identify and confirm. Indigenous groups sometimes deviated, straightened, or prolonged natural streams and rivers. Several examples come to mind (excluding the Casiquiare Canal, where human intervention has not yet been proven), such as the numerous enclosing ditches that control the water level in Guianese raised-field complexes or the features constructed to control runoff around the Hertenrits Mound in western Suriname (Figure 43). The straightening of creeks and rivers by pre-Columbian groups has been little studied in the Amazon, and is better documented in the Andes. In the Cuzco region of Peru, a 200-kilometer stretch of the Vilcanota River and of its tributaries was channeled in segments that together total 23 kilometers of linear earthworks (Farrington 1983). The meanders were moderated through the construction of low retaining walls that controlled irrigation and the erosion of the banks (Deza Rivasplata 2005; Kendall and Rodríguez 2009).

European colonists also modified the rainforest and the coast. Although the polders are easily recognized by their regular geometry and layout, other earthworks are more discrete and ephemeral. During the gold rush in French Guiana that spanned from 1850 to 1900 (Mam Lam Fouck 1999; Montabo 2004), prospectors systematically diverted small creeks to exploit them. Today, it is difficult to distinguish between these canals left over from gold-panning activities and those dug by indigenous groups in the pre-Columbian period.

Nonetheless, it is important to note that the features created by the colonists are less stable and more ephemeral than the pre-Columbian earthworks. Except for the polder canals, which represent a significant investment of labor, the other modifications to the lowlands carried out in the colonial period were much smaller in scale. In less than 50 years, the Creole agricultural beds constructed on the savannas have almost completely disappeared. Those exposed to the open air in the elevated and open zones of the savanna have succumbed to aeolian erosion and strong rainfall. This is particularly clear when observing aerial photographs around Kourou from 1955 to the present. In contrast, the same agricultural beds in the forest are fairly well preserved, whereas it might seem more logical that the tree roots would have destroyed them. A similar phenomenon of gradual destruction affects the features that are called *chinampas* (but which really should be defined as raised fields) in the Xochimilco-Chalco Basins of Central Mexico (Armillas 1971).

In general, raised earthen structures are better preserved in the forested or seasonally flooded environments than in open, dry areas. In contrast, excavated structures tend to be better conserved in the reverse of these conditions.

The problem is similar for raised fields because, although most of the mounds on the Guiana coast are human built, elsewhere in South America there are others that have been created through purely natural processes (Renard et al. 2011). Here again, it is difficult to determine their origin without archaeological, pedological, archaeobotanical, or ecological data. Mounds of natural groups often feature a disconcerting regularity. Ecologists Doyle McKey and Delphine Renard (personal communication, 2010) suspect that the landscapes along the Upper Rio Negro in Colombia that have been previously judged as anthropogenic (Reichel-Dolmatoff and Reichel-Dolmatoff 1974) are, in fact, the result of natural processes. Methods for distinguishing natural and cultural origins of mound-field landscapes are currently being explored (Doyle McKey and Delphine Renard, personal communication, 2011).

It remains to evaluate the efficiency of raised-field agriculture. Is it really a panacea? The spread of these structures across wide areas of numerous South American regions may suggest that this agricultural technique yields exceptional productivity. The few historic accounts also suggest elevated harvests. The yields from the first experimental raised fields 30 years ago were very satisfactory. In Peru in 1990, between 500 and 1,500 hectares of raised fields around Lake Titicaca were cultivated by a few hundred Quechua and Aymara people (Erickson 2006b). Nevertheless, these experiments, followed by others as well as the rehabilitation of ancient raised fields, have shown the limits of the system (Bandy 2005; Erickson 2003b, 2006b; Kolata et al. 1996; Morris 2004; Swartley 2002). In French Guiana in the early 2000s, I encountered some farmers of European origin who had unsuccessfully attempted to use this technique. Clearly, it is prudent to moderate one's enthusiasm regarding the success of this practice. As with other pre-Columbian agricultural methods used in the Amazon, raised fields have a threshold to their efficiency. Indeed, after several years, the returns begin to diminish, and it becomes necessary to construct new structures to replace the older ones (Swartley 2002). Yet, the practice remains a viable one, and the current level of knowledge regarding this technique is greatly inferior to that possessed by the pre-Columbian populations. Thus, despite some reservations, archaeologists who have studied the raised-field method remain convinced of its productivity.[1]

Although direct observation of modern indigenous groups, descendants of the pre-Columbian populations, is essential, it is also necessary to base archaeological interpretations on the field data rather than simply projecting modern phenomena into the past. Indeed, it is futile to attempt to explain the technological systems or social organization of past societies using comparative studies of societies that have been completely transformed by European contact. Other than the physical environment they occupy, modern indigenous communities share little with their pre-Columbian ancestors. Even then, we must question the actual similarity of the environment, recalling that most modern indigenous groups do not inhabit territories of their own choosing, but rather have been forced to relocate by the spread of colonial and modern development. The environmental challenges they face are therefore not the result of poor settlement choices, but rather have been imposed by outside forces. Rather than attempt to understand pre-Columbian lifeways through the prism of direct historical comparison, it is more interesting to try to understand the relations that indigenous peoples have established with the world that they occupy. Descola (2005) defines cross-culturally the principal types of relationships between humans and nonhumans; in particular, he discusses the animistic worldview of Amazonian societies and how it shapes their interactions with the exterior world. To accept this essential indigenous concept of interdependence and continuity between humans and nature is clearly more satisfying and productive scientifically from the perspective of ancient landscapes.

Before the 1980s, it was acceptable among researchers to reject hypotheses that implicated any role of human intervention in the evolution of the Amazonian landscape (e.g., Meggers 1971). Yet, in the last

30 years, a radically opposite idea has developed that pre-Columbian peoples constituted a decisive (perhaps even the principal) influence on the environment. The supporters of this argument often present convincing and well-supported cases (Balée 2010a; Balée and Erickson 2006). Nonetheless, even though the impact of humans was essential, we cannot simply dismiss the effect of nature. Indeed, a middle path can be proposed (Bush and Silman 2007). The alluvial plains, the Andean piedmont, and some of the vast savannas were densely occupied by humans, who played a central and decisive role in their development. Outside these regions, a large part of the Amazonian rainforest was less affected by these transformations. The Amazonian mosaic cannot be considered as a homogeneous whole; rather, it must be considered case by case. If we return to the example of the coastal savannas of Guiana, the powerful impact of pre-Columbian populations on the landscape is well documented (Rostain 1991), but so is the effect of natural processes on this disturbed biotope (McKey et al. 2010). Thus, it is the conjunction of several factors (pre-Columbian earthworks, the careful engineering of earth and water, and the intervention of natural ecosystem engineers) that has produced the forms visible today.

Indigenous people of the Amazon, including Guiana, carried out little construction with stone because of its relative scarcity. Instead, they focused on construction of highly developed earthworks, given the environment and their needs. Archaeologists have long underestimated the degree of skill demonstrated in this pre-Columbian engineering, and were initially completely unaware of its very existence. It is true that the earthworks, often constructed on a densely vegetated landscape, can be difficult to recognize and are easily confused with natural features. While it was generally accepted that these groups had constructed some settlement mounds, these would not have represented a significant investment of labor when considered over a long-term perspective of several centuries. Most interest was focused on works constructed from stone rather than on raw earthen mounds, regardless of how complex and skillfully built. A number of researchers have turned a more nuanced eye to ecofacts (e.g., Denevan 1966, Erickson 1980; Heckenberger 2005; Heckenberger 2009; Nordenskiöld [1916] 2009; Redmond and Spencer 2007; Roosevelt 1991; Walker 2004). What seems obvious today was not until recently. I was met with skepticism when I discovered raised fields in French Guiana 20 years ago. I was told, "The Indians could not have built those. It *must* be the work of slaves or convicts." Ignorant of indigenous cultural heritage, some doubters cannot admit the undeniable superiority of its symbiosis with the Amazonian world. Thus was born a false paradigm of the Amazon that is still difficult to discredit, despite the overwhelming evidence to its contrary that has been emerging over the years.

Finally, a general review of raised-field sites around the world reveals a dearth of scientific studies (Renard et al. 2011). Because they are the only monumental remains of the pre-Columbian populations of the lowlands, these structures have been more carefully studied in Latin America than elsewhere. Furthermore, while the different forms and organizations of the mounds are similar around the world, those of the New World are significantly larger than in other regions (Denevan and Turner II 1974). In any case, most of the research carried out on these types of agricultural sites on five continents demonstrates that they are generally associated with large populations. There are some instances of small communities of farmers cultivating on raised fields (Mollard and Walter 2008), but they are rare and tend to be caused by particular social, political, or economic conditions. In the other instances, raised-field farming is practiced after careful consideration and is associated with a significant population density, perhaps even strong demographic pressure (Denevan and Turner II 1974). A sufficient workforce is necessary to build and maintain this type of agricultural system. There is no longer any doubt about the strong relationship between this type of agriculture and a large population; only the exact calculations of the density for specific cases is in question. Yet, the key question is whether the raised fields allowed the population to grow, or if instead demographic growth required intensification of agricultural techniques. Indeed, the two scenarios seem linked; a high density imposes the need for an intensified production, which itself increases population growth. It is nevertheless important to recall that these communities preferentially chose to settle wetland environments, which implies a preexisting need for abundant resources and, above all, a close interaction between humans and their environment.

Moreover, the use of this technique affects the society. Whereas a small group such as the Palikur from the north of Amapá in the eighteenth century could build a small number of mounds, the intensification of the system engenders socioeconomic adjustments within the community. When the spatial extent of cultivation increases, supervision of the labor by a specialist could become necessary, but not always (Erickson 2006a). Thus, some individuals dedicate themselves more to farming at the expense of other activities that must now be delegated to other members of the community. Specialization begins to develop. With the production of surplus, exchange with communities in the interior zone becomes an important possibility. In this way, the transformation of the landscape is closely linked with structural changes to the society. The fisher-farmers of the coast of Bénin offer an excellent example of this coupled process. In this region, it has been demonstrated to be the fundamental role of social change, religious influence, and pressure from neighboring groups on the modification of the environment (Leite et al. 2003). This close link between society and agricultural earthworks exists in many regions around the world (Mollard and Walter 2008).

For a dozen years now, high-profile publications by Amazonian archaeologists in journals such as *Science, Nature,* and *Proceedings of the National Academy of Science* have increased, bringing to light new discoveries, each more impressive than the last. First there were the earthworks of the Llanos of Beni, Bolivia (Erickson 2000a), the intense transformations of the Upper Xingu landscape (Heckenberger et al. 2003), the ancient impact of humans on the rainforest (Willis et al. 2004), the large territories and modifications of the Upper Xingu (Heckenberger 2009; Heckenberger et al. 2008), and the raised fields of French Guiana (Iriarte et al. 2012; McKey et al. 2010). Not simply the trend of the moment, this profusion of exceptional discoveries reflects the real progress and growth of archaeology in the Amazon.

The ecofacts produced by the intentional modification of the landscape for agriculture are among the most representative remains. At first glance ordinary and simple, they turn out to be impressive and rich in information when examined with a more complex approach. The goal of this book was to demonstrate that only by combining data from a variety of disciplines can we hope to comprehensively understand anthropogenic landscapes. A multiplicity of perspectives allows a more nuanced understanding of environments with complex origins. Such work calls for archaeobotany, archaeology, biology, botany, ecology, ethnohistory, ethnology, geography, geomorphology, history, hydrology, pedology, physics, and remote sensing. Few maps of raised fields exist for other regions of South America that are as precise and complete those for the coast of Guiana. These latter provide an image that is likely very close to the reality of the original pre-Columbian landscape. This cartography sometimes remains very general, with large "blobs" marking modified zones, or, in other regions, may be very precise, indicating the boundaries of specific complexes. Mapping of the ancient anthropogenic impact on the coast of Guiana has been coupled with interdisciplinary research that has allowed us to better specify the research questions, the data collection methods, the cognitive paths toward interpretation, and the results obtained. We can only understand the ancient anthropization of the landscape by disentangling multiple strands of evidence, which is a difficult, if not impossible, task for the archaeologist to undertake alone.

Notes

Foreword

1 By contrast, the Achuar have a mean population density of fewer than two inhabitants per 10 square kilometers.

2 For the Vaupés, see Jackson (1983); for the Xingu, see Franchetto and Heckenberger (2001).

Introduction

1 ("*Qued'eau!*") Exclaimed in 1857 by Patrice de Mac Mahon, President of France, while visiting a city in the south of France that had been flooded by the Garonne River.

2 A word originating from Nahuatl that means "place enclosed by canes" (*chinámitl* = enclosed by canes; *pan* = place).

3 "*Hortillonnage*" is derived from the Picard term *hortillon,* itself derived from the Latin *hortellus*, meaning "little garden." These are drained parcels separated by canals, located near Amiens, and used for market production. Used since the Gallo-Roman period, today only 300 hectares of the original 10,000 remain. They currently support 1,000 people (Mollard and Walter 2008).

4 For this reason, chinampas are sometimes incorrectly referred to as "floating gardens."

5 The *marombas* of the Amazon (moored rafts) also host herds of cattle during the season of high waters.

Chapter 3

1 In Amazonian archaeology, "splitters" (who multiply the number of ceramic types) tend to outnumber the "lumpers" (who take on a more global perspective and tend to combine types).

Chapter 4

1 In contrast, a "*portage*" is a path that cuts through a bend in the river or the rapids, designed to transport the canoe.

2 The "*nivrée*" is fishing with poison using macerated plant material with bioactive principles that paralyze the gills of fish, causing them to rise to the surface clumsily, from several vines in the Sapindaceae (soapberry) and Fabaceae (bean) families or from two cultivated shrubs, one in the Fabaceae family and the other in the Asteraceae (sunflower) family (William Balée, personal communication, 2012).

CHAPTER 5

1 It is amusing to note a certain similarity between French Guiana and New York City. First, Hudson Bay also was christened "New Angoulême" when it was discovered in 1524. Next, thanks to (or because of) the Treaty of Breda in 1667, the Dutch exchanged New Amsterdam, the settlement of Manhattan Island, for their sovereignty in Suriname (Weil 2005). The twin nature of New York/Guiana stops there, with each of these places subsequently leading quite distinct histories.

2 The prison system undertook this measure to prevent the convicts from escaping by using logs to build rafts. When observing the dense growth of coconut trees today, it is difficult to imagine what these three islands must have looked like when they were totally cleared a few decades ago.

3 This unsigned manuscript transcribes the moving recollections of 22 persons who lived during the "pre-base" period. This rare document, written and compiled in 1999, unfortunately has never been published.

CONCLUSION

1 Renard et al. (2011) provide a detailed synthesis of what is known (and not known) about raised-field agroecology.

Bibliography

Abbott, M. B., M. W. Binford, M. Brenner, and K. R. Kelts. 1997. "A 3500 ^{14}C yr high-resolution record of water-level changes in lake Titicaca, Bolivia/Peru." *Quaternary Research* 47:169–180.

Abonnenc, E., Y. Le Lay, and H. Lecoq. 1956. "Démographie de la Guyane française, III: Les indiens Galibis." *Journal de la Société des Américanistes* 45:195–208.

Adam, D. 1936. *El Dorado: La Guyane Française Agricole*. Paris: Larose.

Adams, R. E. W. 1982. "Les Canaux Mayas." *La Recherche* 13(136):1066–1067.

Ahlbrinck, W. 1931. *Encyclopaedie der Karaïben*. Verh. Kon. Ak. von Wetenschappen, Nieuwe Reeks deel XXVII(1). Amsterdam: Koniklijke Akademie van Wetenschappen.

Alès, C., and M. Pouyllau. 1992. "La Conquête de l'Inutile: Les Géographies Imaginaires de L'Eldorado." *L'Homme*, 122–124, "La Redécouverte de l'Amérique," Paris: EHESS, pp. 271–308.

Anonymous. 2000. Je me Souviens . . . Kourou. Korou.

Antczak, M. M., and A. Antczak. 2006. *Los Idolos de la Islas Prometidas: Arqueología Prehispánica del Archipiélago de Los Roques*. Caracas: Editorial Equinoccio/Universidad Simón Bolivar.

Arce, J. 1993. "Evaluación y Comparación de Rendimientos de Cuatro Cultivos en Tres Anchuras de Camellones (Campos Elevados) en la Estación Biológica del Beni (Prov. Ballivián, Dpto. Beni)" [Ph.D. dissertation]. Department of Agronomy, Universidad Técnica del Beni, Trinidad.

Armillas, P. 1971. "Gardens on Swamps." *Science* 174(4010):653–661.

Arellano Moreno, A. 1964. *Relaciones Geográficas de Venezuela*. Caracas: Academia Nacional de la Historia.

Arroyo-Kalin, M. 2008. "Steps Towards an Ecology of Landscape: A Geoarchaeological Approach to the Study of Anthropogenic Dark Earths in the Central Amazon Region" [Ph.D. dissertation]. Department of Archaeology, University of Cambridge, United Kingdom.

———. 2010. "The Amazonian Formative: Crop Domestication and Anthropogenic Soils." *Diversity* 2: 473–504. Available online at www.mdpi.com/journal/diversity.

———. 2012. "Slash-burn-and-churn: Landscape History and Crop Cultivation in pre-Columbian Amazonia." *Quaternary International* 249:4–18.

Arvelo, B. L. 1978. "Tucuragua, un Yacimiento Arauquinoide del Orinoco Medio." In *Proceedings of the 7th International Congress for the Study of Pre-Columbian Cultures of the Lesser Antilles*, Caracas, pp. 29–37.

Arvelo-Jiménez, N., and H. Biord. 1994. "The Impact of Conquest on Contemporary Indigenous Peoples of the Guiana Shield: The System of Orinoco Regional Interdependence." In *Amazonian Indians from Prehistory to the Present: Anthropological Perspectives*, edited by A. C. Roosevelt, pp. 55–78. Tucson: University Arizona Press.

Atlas de Guyane. 1979. *Atlas des Départements d'Outre-Mer, No. IV: La Guyane*. Bordeaux-Talence: CEGET-CNRS/ORSTOM.

Aujoulat, N., J. Briand, and P. Grenand (eds.). 1997. *L'Archéologie en Guyane*. Cayenne: Édition APPAAG.

Balée, W. 1987. "Cultural Forest of the Amazon." *Garden* 11(6):12–14.

———. 1989. "The Culture of Amazonian Forests." In *Ressource Management in Amazonia: Indigenous and Folk Strategies*, edited by D. A. Posey and W. Balée, pp. 1–21. New York: Advances in Economic Botany, Vol. 7.

———. (ed.). 2010a. "Long-term Anthropic Influences on the Diversity of Amazonian Landscapes and Biota." *Diversity* Special Issue, available at www.mdpi.com/journal/diversity.

———. 2010b. "Amazonian Dark Earths." *Tipití*, 8:paper 3.

Balée, W., and C. L. Erickson (eds.). 2006. *Time and Complexity in Historical Ecology*. New York: Columbia University Press.

Bandy, M. S. 2005. "Energetic Efficiency and Political Expediency in Titicaca Basin Raised Field Agriculture." *Journal of Anthropological Archaeology* 24(3):271–296.

Barba, J., E. Canal, E. García, E. Jordà, M. Miró, E. Pastó, R. Playà, I. Romero, M. Via, and E. Woynarovich. 2003. *Moxos: Una Limnocultura. Cultura y Medio Natural en la Amazonia Boliviana*. Barcelona: CEAM.

Barlow, J., T. A. Gardner, A. C. Lees, L. Parry, and C. A. Peres. 2011. "How Pristine Are Tropical Forests? An Ecological Perspective on the pre-Columbian Human Footprint in Amazonia and Implications for Contemporary Conservation." *Biological Conservation*. doi:10.1016/j.biocon.2011.11.011.

Barone-Visigalli, E., M.-T. Prost, and S. Rostain. 1991. *Modalités d'Occupation des Sites Amérindiens en Guyane: le Cas de La Sablière, Kourou*. Cayenne: Rapport à la Direction des Antiquités.

Barre, L. de la. 1666. *Description de la France Equinoxiale* (Lk12 789). Paris: Bibliothèque Nationale.

Barrère, P. 1743. *Nouvelle Relation de la France Equinoxiale*. Paris: Piget-Damonneville, Durand.

Barse, W. P. 1990. "Preceramic Occupations on the Orinoco River Valley." *Science* 250:1388–1390.

———. 1995. "El Periodo Arcaico en el Orinoco y su Contexto en el Norte de Sur America." In *Ambito y Ocupaciones Tempranas de la América Tropical*, edited by I. Cavelier and S. Mora, pp. 99–114. Santa Fe de Bogotá: Fundación ERIGAIE/Instituto Colombiano de Antropología.

Biet, A. 1664. *Voyage dans la France Equinoxiale en l'Isle de Cayenne Entrepris par les Français en l'Année 1652*. Paris: Clouzier Éditeur.

Binford, M. W., A. L. Kolata, M. Brenner, J. W. Janusek, M. T. Seddon, M. Abbott, and J. H. Curtis. 1997. "Climate Variation and the Rise and Fall of an Andean Civilization." *Quaternary Research* 47:235–248.

Boada Rivas, A. M. 2006. *Patrones de Asentamiento Regional y Sistemas de Agricultura Intensiva en Cota y Suba, Sabana de Bogotá (Colombia)*. Bogotá: Fundación de Investigaciones Arqueológicas Nacionales/Banco de la República.

Boomert, A. 1976. "Pre-Columbian Raised Fields in Coastal Surinam." In *Proceedings of the 6th International Congress for the Study of the Pre-Columbian Cultures of the Lesser Antilles*, pp. 134–144. Gainesville, Florida.

———. 1978. "Prehistoric Habitation Mounds in the Canje River Area?" *Journal of the Walter Roth Museum of Archaeology and Anthropology* 1(1):44–51.

Boomert, A. 1979. "The Prehistoric Stone Axes of the Guianas: A Typological Classification." *Journal of the Walter Roth Museum of Archaeology and Anthropology* 2(2):99–124.

———. 1980a. "The Sipaliwini Archaeological Complex of Surinam: A Summary." In *Nieuve West-A. Gids*, Vol. 2, pp. 94–107. Utrecht: Stichting Nieuwe West-Indische Gids.

———. 1980b. "Hertenrits: An Arauquinoid Complex in North West Suriname." *Journal of the Walter Roth Museum of Archaeology and Anthropology* 3(2):68–104.

———. 1981. "The Taruma Phase of Southern Suriname." *Journal of the Walter Roth Museum of Archaeology and Anthropology* 4(1–2):104–157.

———. 1983. "The Saladoid Occupation of Wonotobo Falls, Western Surinam." In *Compte Rendu du 9e Congrès International d'Études des Civilisations Précolombiennes des Petites Antilles*, pp. 97–120. Montréal: Université de Montréal Press.

———. 1986. "The Cayo Complex of St Vincent: Ethnohistorical and Archaeological Aspects of the Island Carib Problem." *Antropológica* 66:3–68.

———. 1987. "Gifts of the Amazons: 'Green Stone' Pendants and Beads as Items of Ceremonial Exchange in Amazonia and the Caribbean." *Antropológica* 67:33–54.

———. 1993. "The Barbakoeba Archaeological Complex of Northeast Suriname." *OSO: Tijdschrift voor Surinaamse Taalkunde, Letterkunde, Cultuur en Geschiedenis* 12(2):198–222.

———. 2004. "Koriabo and the Polychrome Tradition: The Late-prehistoric Era between the Orinoco and Amazon Mouths." In *Late Ceramic Societies in the Eastern Caribbean*, edited by A. Delpuech and C. Hofman, pp. 251–268. Paris: BAR 1273, Monographs in American Archaeology 14.

Boomert, A., and S. B. Kroonenberg. 1977. "Manufacture and Trade of Stone Artifacts in Prehistoric Surinam." In *Ex Horreo, I.P.P. 1951–1976: Cingula IV*, edited by Van Beek, Brandt, and Groenman-van Waateringe, pp. 9–46. Amsterdam: Universiteit van Amsterdam.

Bouyer, F. 1867. *La Guyane Française: Notes et Souvenirs d'un Voyage Exécuté en 1862-1863*. Paris: Librairie Hachette.

Bray, W. 1995. "Searching for Environmental Stress: Climatic and Anthropogenic Influences on the Landscape of Colombia." In *Archaeology in the Lowland American Tropics*, edited by P. W. Stahl, pp. 96–112. Cambridge: Cambridge University Press.

Bray, W., L. Herrera, M. Cardale Schrimpff, P. Botero, and J. G. Mosalve. 1987. "The Ancient Agricultural Landscape of Calima, Colombia." In *Prehispanic Agricultural Fields in the Andean Region*, edited by W. Denevan, K. Mathewson, and G. Knapp, pp. 443–481. BAR International Series No. 359. Oxford: Archaeopress.

Broadbent, S. M. 1964. "Agricultural Terraces in Chibcha Territory, Colombia." *American Antiquity* 29(4):501–504.

———. 1987. "The Chibcha Raised-field System in the Sabana de Bogotá, Colombia: Further Investigations." In *Prehispanic Agricultural Fields in the Andean Region*, edited by W. Denevan, K. Mathewson, and G. Knapp, pp. 425–442. BAR International Series No. 359. Oxford: Archaeopress.

Buol, S. W., R. J. Southard, R. C. Graham, and P. A. McDaniel. 2003. *Soil Genesis and Classification*. Ames: Iowa State Press.

Bush, M. B., D. R. Piperno, and P. A. Colinvaux. 1989. "A 6,000-year History of Amazon Maize Cultivation." *Nature* 340:103–305.

Bush, M. B., and M. R. Silman. 2007. "Amazonian Exploitation Revisited: Ecological Asymmetry and the Policy Pendulum." *Frontier in Ecology and Environment* 5(9):457–465.

Bush, M. B., M. R. Silman, C. McMichael, and S. Saatchi. 2008. "Fire, Climate Change and Biodiversity in Amazonia: A Late-Holocene Perspective." *Philosophical Transactions of the Royal Society* 363:1795–1802.

Butt Colson, A. 1973. "Inter-tribal Trade in the Guiana Highlands." *Antropológica* 34:5–70.

Cadudal, F. 2007. "Camellones et Systèmes Agricoles Préhispaniques des Basses Terres de la Côte Nord de l'Équateur." Available online at www.arqueo-ecuatoriana.ec.

Caillavet, C. 1983. "Toponimia Histórica y Formas Prehispánicas de Agricultura en la Région de Otavalo, Ecuador." *Bulletin de l'Institut Français d'Études Andines* 12(3–4):1–21.

———. 2006. "Historia y Agricultura Autoctona en los Andes Ecuatorianos: el Complejo Campos Elevados, en Ecosistemas Diversos (Siglos XV–XVII)." In *Agricultura Ancestral: Camellones y Albarradas*, edited by F. Valdez, pp. 111–126. Quito: Abya-Yala.

Carneiro, R. L. 1979. "Tree Felling with a Stone Axe: An Experiment Carried Out among the Yanomamö Indians of Southern Venezuela." In *Ethnoarchaeology: Implications of Ethnography for Archaeology*, edited by C. Kramer, pp. 21–58. New York: Columbia University Press.

Cavelier, I. C., C. Rodriguez, L. F. Herrera, G. Morcote, and S. Mora. 1995. "No Solo de Caza Vive el Hombre: Ocupación del Bosque Amazónico Holoceno Temprano." In *Ambito y Ocupaciones Tempranas de la América Tropical*, edited by I. Cavelier and S. Mora, pp. 27–44. Santa Fé de Bogotá: Fundación Erigaie-Instituto Colombiano de Antropologia.

Chabrillan, M. de. 1743. *Journal des Observations que j'ai Fait et de ce qui s'est Passé Pendant les Années 1742 et 1743*. Archives de la Marine, France.

Charcornac-Rault, M., and S. Rostain. In Press. "Les Plantes Cultivées dans les Champs Surélevés Précolombiens." In *Archéologie en Amazonie Française: Préhistoire du Littoral Occidental de Guyane*, edited by S. Rostain. BAR International Series, Paris Monographs in American Archeology. Oxford: Archeopress.

Chantre y Herrera, J. (1767) 1901. *Historia de las Misiones de la Compañía de Jesús en el Marañon Español, 1637–1767*. Madrid.

Charles-Dominique, P., M. L. Absy, M.-P. Ledru, B. Riéra, C. Sardou, M. Servant, and B. Turcq. 1996. "Incidents Climatiques et Modifications de l'Écosystème depuis 3000 Ans en Guyane Française et en Amérique du Sud." In *Dynamique à Long Terme des Écosystèmes Forestiers Intertropicaux: Résumés*, pp. 223–228. Paris: ORSTOM.

Charles-Dominique, P., P. Blanc, D. Larpin, M.-P. Ledru, B. Riera, C. Sarthou, M. Servant, and C. Tardy. 1998. "Forest Perturbations and Biodiversity during the Last Ten Thousand Years in French Guiana." *Acta Oecologica, International Journal of Ecology* 19:295–302.

Chrétien, J. 1719. *Lettres d'un Missionnaire de la Compagnie de Jésus à un Père de la Même Compagnie Écrites de Cayenne en Amérique*. Manuscript 406 from La Bibliothèque Méjane, Aix-en-Provence.

Clark, K., and C. Uhl. 1987. "Farming, Fishing and Fire in the History of the Upper Negro Region of Venezuela." *Human Ecology* 15:1–26.

Clerc, L., and L. Déodat. 2009. *Rapport de Prospections Kourou 2009*. Paris: CNRS/UMR8096/Université Paris 1.

Clément, C. 1999a. "1492 and the Loss of Amazonian Crop Genetic Resources, I: The Relation between Domestication and Human Population Decline." *Economic Botany* 53:188–202.

———. 1999b. "1492 and the Loss of Amazonian Crop Genetic Resources, II: Crop Biogeography at Contact." *Economic Botany* 53:203–216.

Clément, C., R. M. de Cristo-Araújo, G. Coppens d'Eeckenbrugge, A. Alves Pereira, and D. Picanço-Rodrigues. 2010. "Origin and Domestication of Native Amazonian Crops." *Diversity* 2:72–106. Available at www.mdpi.com/journal/diversity.

Colomb, C. 1992. *Œuvres Complètes*. Paris: La Différence.

Colinvaux, P. A. 1989. "The Past and Future Amazon." *Scientific American* 260:102–108.

Colinvaux, P. A., M. C. Miller, L. Kam-Biu, M. Steinitz-Kannou, and I. Frost. 1985. "Discovery of Permanent Amazon Lakes and Hydraulic Disturbance in the Upper Amazon Basin." *Nature* 313:12–15.

Cooke, R. G., and D. Piperno. 1996. "Le Peuplement de l'Amérique Central et de l'Amérique du Sud et les Adaptations aux Forêts Tropicales avant la Colonisation Européenne." In *L'Alimentation en Forêt Tropicale: Interactions Bioculturelles et Perspectives de Développement, Vol. 1, Les Resources Alimentaires: Production et Consommation*, edited by C. M. Hladik et al., pp. 77–96. Paris: UNESCO.

Coréal, F. de. 1722. *Voyages de François Coreal aux Indes Occidentales, 1666–1697*, edited by A. Cailleau. Paris.

Cornette, A. 1987. "Quelques Données sur l'Occupation Amérindienne dans la Région Basse Mana, Bas Maroni, d'après les Sources Ethno-archéologiques." *Equinoxe* 24:70–99.

Cornette, A., P. Nowacki-Breczwski, O. Puaux, and E. Visigalli. 1992. "Archéologie et Géomorphologie: Quelques Exemples d'Études des Sites Littoraux." In *Évolution des Littoraux de Guyane et de la Zone Caraïbe Méridionale pendant le Quaternaire*, pp. 93–103. Colloques et Séminaires. Paris: Éditions de l'ORSTOM.

Costa, F. W. 2009. *Arqueologia das Campinaranas do Baixo Rio Negro: em Busca dos Pré-ceramistas no Areais da Amazônia Central* [Ph.D. dissertation]. Programa de Pós-graduação em Arqueologia, Universidade de São Paulo, Brazil.

Coudreau, H. A. 1891. "Chez nos Indiens: Quatre Années dans la Guyane Française (1887–1891)." *Le Tour du Monde* LXIII:1–96.

Coutet, C. 2010. *La Céramique Amérindienne du Littoral de Guyane: Ruptures pour Quelles Continuités?* [Ph.D. dissertation]. Université de Paris-1, Paris.

Cremers, G. 1990. *Petite Flore Illustrée, les Savanes Côtières*. Nature Guyanaise No. 5–6. Cayenne : Sepanguy/ORSTOM.

Crevaux, J. 1882. "Voyage d'Exploration à Travers la Nouvelle-Grenade et le Vénézuela: Rio Magdalena, Rio Guaviare, Orinoco." Paris: Le Tour du Monde.

Cruxent, J. M. 1972. "Tupuquén: un Yacimiento con Lítica de Tipo Paleo-Indio." *Acta Cientifica Venezolana* 23:1–17.

Cruxent, J. M., and I. Rouse. 1958–59. *An Archaeological Chronology of Venezuela*. Social Science Monography No. 6. Washington, D.C.: Pan American Union.

Daly, D. C., and J. D. Mitchell. 2000. "Lowland Vegetation of Tropical South America: An Overview." In *Imperfect Balance: Landscape Transformations in the Pre-Columbian Americas*, edited by D. Lentz, pp. 391–454. New York: Columbia University Press.

Darch, J. P. (ed.). 1983. *Drained Field Agriculture in Central and South America*. BAR International Series No. 189. Oxford: Archaeopress.

Delawarde, J.-B. 1967. "Les Galibis de la Mana et d'Iracoubo, Guyane." *Journal de la Société des Américanistes* 55(2):333–388.

Delpech, S. 2005. *Carrière des Ananas (Saint-Laurent du Maroni)*. Rapport de Diagnostic. Cayenne: INRAP.

DeMenocal, P. B. 2001. "Cultural Responses to Climate Change during the Late Holocene." *Science* 292:667–673.

Denevan, W. M. 1963. "Additional Comments on the Earthworks of Mojos in Northeastern Bolivia." *American Antiquity* 28(4):540–545.

———. 1966. *The Aboriginal Cultural Geography of the Llanos de Mojos of Bolivia*. Ibero-America No. 48. Berkeley/Los Angeles: University of California Press.

———. 1970. "Aboriginal Drained-field Cultivation in the Americas." *Science* 169:647–654.

———. 1992a. "Native American Populations in 1492: Recent Research and a Revisited Hemispheric Estimate." In *The Native Population of the Americas in 1492*, edited by W. Denevan, pp. xvii–xxxviii. Madison: University of Wisconsin Press.

———. 1992b. "Stone vs. Metal Axes: The Ambiguity of Shifting Cultivation in Prehistoric Amazonia." *Journal of the Steward Anthropological Society* 20:153–165.

———. 1996. "A Bluff Model of Riverine Settlement in Prehistoric Amazonia." *Annals of the Association of American Geographers* 86(4):654–681.

———. 2001. *Cultivated Landscapes of Native Amazonia and the Andes*. New York: Oxford University Press.

———. 2004. "Semi-intensive pre-European Cultivation and the Origins of Anthropogenic Dark Earths in Amazonia." In *Amazonian Dark Earths: Explorations in Space and Time*, edited by B. Glaser and W. I. Woods, pp. 135–143. Berlin/London: Springer.

Denevan, W. M., and K. Mathewson. 1983. "Preliminary Results of the Sambirondón Raised-field Project, Guyayas Basin, Ecuador." In *Drained Field Agriculture in Central and South America*, edited by J. P. Darch, pp. 167–181. BAR International Series No. 189. Oxford: Archaeopress.

Denevan, W. M., K. Mathewson, and G. Knapp (eds.). 1987. *Prehispanic Agricultural Fields in the Andean Region*. BAR International Series No. 359. Oxford: Archaeopress.

Denevan, W. M., and K. H. Schwerin. 1978. "Adaptive Strategies in Karinya Subsistence, Venezuelan Llanos." *Antropológica* 50:3–91.

Denevan, W. M., and B. L. Turner II. 1974. "Forms, Functions and Associations of Raised Fields in the Old World Tropics." *The Journal of Tropical Geography* 39:24–33.

Denham, T. P. 2005. "Les Racines de l'Agriculture en Nouvelle-Guinée." *La Recherche* 389:48–51.

Denham, T. P., S. G. Haberle, C. Lentfer, R. Fullagar, J. Field, M. Therin, N. Porch, and B. Winsborough. 2003. "Origins of Agriculture at Kuk Swamp in the Highlands of New Guinea." *Science* 301(5630):189–193.

Denham, T. P., S. Haberle, and C. Lentfe. 2004. "New Evidence and Revised Interpretations of Early Agriculture in Highland New Guinea." *Antiquity* 78(302):839–857.

Derby, O. 1879. "The Artificial Mounds of the Island of Marajo." *American Naturalist* 13:224–229.

———. 1898. "A Ilha de Marajó." *Revista do Instituto Histórico, Geográfico e Ethnográfico do Pará* 2:163–273.

Descola, P. 1986. *La Nature Domestique: Symbolisme et Praxis dans l'Écologie des Achuar.* Paris: Fondation Singer-Polignac/Éditions de la Maison des Sciences de l'Homme.

———. 1993. *Les Lances du Crépuscule: Relations Jivaros, Haute Amazonie.* Collection Terre Humaine. Paris: Plon.

———. 1994. *In the Society of Nature.* Cambridge University Press, United Kingdom.

———. 2005. *Par-delà Nature et Culture.* Paris: Nrf, Gallimard.

Dillehay, T. D., M. Pino Quivira, R. Bonzani, C. Silva, J. Wallner, and C. Le Quesne. 2007. "Cultivated Wetlands and Emerging Complexity in South-central Chile and Long Distance Effects of Climate Change." *Antiquity* 81:949–960.

Deza Rivasplata, J. 2005. *El Agua de los Incas.* Lima: Universidad Alas Peruanas.

Dotte, E., C. Sand, J. Bolé, and A. Ouetcho. 2005. "Processus de Territorialisation dans l'Espace Archéogéographique de la Nouvelle-Calédonie Pré-contact: le Cas du Nord de la Grande Terre." *Actes du Colloque Corail,* Nouméa.

Drake, F. 1996. *Histoire Naturelle des Indes : The Drake Manuscript in the Pierpont Morgan Library.* New York: W. W. Norton & Company.

Dreyfus, S. 1981. "Notes sur la Chefferie Taino d'Aiti: Capacités Productrices, Ressources Alimentaires, Pouvoirs dans une Société Précolombienne de Forêt Tropicale." *Journal de la Société des Américanistes* 64:229–248.

———. 1992. "Les Réseaux Politiques Indigènes en Guyane Occidentale et leurs Transformations aux XVIIe et XVIIIe Siècles." *L'Homme* 32(122–124):75–98.

———. 1993. "Et Christophe Colomb Vint . . . De la Découverte à la Disparition des Indigènes des Îles." *Ethnies: Chroniques d'une Conquête* 7(14):94–106.

Dubois, M., and C. Guy. 1897. *Album Géographique, Tome 2: les Régions Tropicales.* Paris.

Duputie, A., P. David, C. Debain, and D. McKey. 2007. "Natural Hybridization between a Clonally Propagated Crop, Cassava (*Manihot esculenta* Crantz) and a Wild Relative in French Guiana." *Molecular Ecology* 16:3025–3038.

Durand, J.-M. 2010. *L'Autre Histoire de la Nouvelle-Angoulême: Une Utopie Agricole dans la Guyane du XIXe Siècle.* Collection Espace Outre-mer. Matoury: Ibis Rouge Éditions.

Eden, M., W. Bray, L. Herrera, and C. McEwan. 1984. "Terra Preta Soils and Their Archaeological Context in the Caquetá Basin of Southeast Colombia." *American Antiquity* 49(1):125–140.

Erickson, C. L. 1980. "Sistemas Agrícolas Prehispánicos en los Llanos de Mojos." *America Indígena* 40(4):731–755.

———. 1987. "The Dating of Raised-field Agriculture in the Lake Titicaca Basin, Peru." *In Pre-Hispanic Agricultural Fields in the Andean Region,* Part 1, edited by W. Denevan, D. Mathewson, and G. Knapp, pp. 373–384. BAR International Series No. 359(i). Oxford: Archaeopress.

———. 1993. "The Social Organization of Prehispanic Raised Field Agriculture in the Lake Titicaca Basin." In *Economic Aspects of Water Management in the Prehispanic New World,* edited by V. L. Scarborough and B. L. Isaac, pp. 7369–7426. Research in Economic Anthropology, Supplement 7. Greenwich: JAI Press.

———. 1995. "Archaeological Methods for the Study of Ancient Landscapes of the Llanos de Mojos in the Bolivian Amazon." In *Archaeology in the Lowland American Tropics: Current Analytical Methods and Applications,* edited by P. W. Stahl, pp. 66–95. Cambridge: Cambridge University Press.

———. 2000a. "An Artificial Landscape-scale Fishery in the Bolivian Amazon." *Nature* 408:190–193.

Erickson, C. L. 2000b. "The Lake Titicaca Basin: A Precolumbian Built Landscape." In *Imperfect Balance: Landscape Transformations in the Precolumbian Americas*, edited by XXX Lentz, pp. 311–356. New York: Columbia University Press.

———. 2000c. "Lomas de Ocupación en los Llanos de Moxos." In *Arqueología de las Tierras Bajas*, edited by A. Durán Coirolo and R. Bracco Boksar, pp. 207–226. Montevideo: Comision Nacional de Arqueologia, Ministerio de Educacion y Cultura.

———. 2003a. "Historical Ecology and Future Explorations." In *Amazonian Dark Earths: Origin, Properties, Management*, edited by J. Lehmann et al., pp. 455–500. Dordrecht: Kluwer Academic Publishers.

———. 2003b. "Agricultural Landscapes as World Heritage: Raised Field Agriculture in Bolivia and Peru." In *Managing Change: Sustainable Approaches to the Conservation of the Built Environment*, edited by J.-M. Teutonico and F. Matero, pp. 181–204. Los Angeles: Getty Conservation Institute.

———. 2006a. "Intensification, Political Economy, and the Farming Community in Defense of a Bottom-up Perspective of the Past." In *Agricultural Strategies*, edited by J. Marcus and C. Stranish, pp. 334–363. Los Angeles: Cotsen Institute of Archaeology, University of California.

———. 2006b. "El Valor Actual de los Camellones de Cultivo Precolombinos: Experiencias del Perú y Bolivia." In *Agricultura Ancestral. Camellones y Albarradas: Contexto Social, Usos y Retos del Pasado y del Presente*, edited by F. Valdez, pp. 315–339. Quito: Ediciones Abya-Yala.

———. 2008. "Amazonia: The Historical Ecology of a Domesticated Landscape." In *Handbook of South American Archaeology*, edited by H. Silverman and W. Isbell, pp. 157–183. New York: Springer/Kluwer/Plenum.

Erickson, C., P. Álvarez, and S. Calla. 2008. *Zanjas Circundantes: Obras de Tierra Monumentales de Baures em la Amazônia Boliviana*. Beni: Projecto Agro-Arqueologico Del Beni.

Evans, C., and B. J. Meggers. 1960. *Archeological Investigations in British Guiana*. Bureau of American Ethnology Bulletin 177. Washington, D. C.: Smithsonian Institution.

Farrington, I. S. 1983. "Prehistoric Intensive Agriculture: Preliminary Notes on River Canalization in the Sacred Valley of the Incas." In *Drained Field Agriculture in Central and South America*, edited by J. P. Darch, pp. 221–235. BAR International Series No. 189. Oxford: Archeopress.

Figuti, L. 1992. *Les Sambaqui Copisa (4200 à 12000 ans B.P.): Étude de la Subsistance Chez les Peuples Préhistoriques de Pêcheurs-ramasseurs de Bivalves de la Côte Centrale de l'Etat de São Paulo, Brésil* [Ph.D. dissertation]. Muséum National d'Histoire Naturelle, Paris.

Fisher, R. F., C. Newell Bourn, and W. F. Fisher. 1995. "Opal Phytoliths as an Indicator of the Floristics of Prehistoric Grasslands." *Geoderma* 68:243–255.

Fleury, M. 2000. "L'Agriculture Wayana: Une Transition Viable." In *Les Peuples des Forêts Tropicales Aujourd'hui, Vol. 4: Tégion Caraïbes (Guyane, Belize)*, edited by P. Grenand, pp. 151–161. Brussels: APFT, ULB.

Folgarait, P. J., F. Thomas, T. Desjardins, M. Grimaldi, P. Curmi, I. Tayasu, and P. M. Lavelle. 2003. "Soil Properties and Macrofauna Community in Recently Abandoned Irrigated Rice Fields in Northeastern Argentina." *Biology and Fertility of Soils* 38:358–366.

Forest, J. de. 1625. *Manuscrit d'un Journal de Voyage Fait par les Pères de Familles Envoyés par Mrs les Directeurs de la Compagnie des Indes Occidentales pour*

Visiter la Coste de Guiane: Avec Plusieurs Remarques Curieuses et Quantité de Planches Très Particulières. Sloane Manuscript 179b, British Library Department of Manuscripts, London, United Kingdom.

Franchetto, B., and M. Heckenberger (eds.). 2001. *Os Povos do Alto Xingu—História e Cultura*. Rio de Janeiro: UFRJ.

Fraser, J. A. 2010. "The Diversity of Bitter Manioc (*Manihot Esculenta* Crantz) Cultivation in a Whitewater Amazonian Landscape." *Diversity* 2:586–609. Available at www.mdpi.com/journal/diversity.

Fraser, J. A., and C. R. Clément. 2008. "Dark Earths and Manioc Cultivation in Central Amazonia: A Window on Pre-Colombian Agricultural Systems?" *Boletim do Museu Paraense Emílio Goeldi. Ciências Humanas* 3:175–194.

Frenzel, B. 1979. "L'Homme comme Facteur Géologique en Europe." *Bulletin de l'Association Française pour l'Étude du Quaternaire* 4:191–199.

Fretey, J. 2002. "Les Tortues Marines." In *Guyane ou le Voyage Écologique*, edited by C. Richard-Hansen and R. Le Guen, pp. 224–239. Garies: Éditions Roger Le Guen.

Garson, A. G. 1980. *Prehistory, Settlement and Food Production in the Savanna Region of La Calzada de Paez, Venezuela* [Ph.D. dissertation]. Department of Anthropology, Yale University, New Haven, Connecticut.

Gassón, R. A. 1998. *Prehispanic Intensive Agriculture, Settlement Pattern and Political Economy in the Western Venezuelan Llanos* [Ph.D. dissertation]. Department of Anthropology, University of Pittsburgh, Pennsylvania.

Geijskes, D. C. 1964. "Les Arawaks dans la Préhistoire de Surinam." In *Comptes Rendus du 1er Congrès International d'Études des Civilisations Précolombiennes des Petites Antilles*, pp. 71–83. Fort-de-France, Martinique.

Gely, A. 1984. "L'Agriculture sur Brûlis Chez quelques Communautés d'Amérindiens et de Noirs Réfugiés de Guyane française." *Journal d'Agriculture Traditionelle et de Botanique Appliquée* 31(1–2):43–70.

Georgescu, P. C., C. P. Constantino Georgescu, and A. Laraque. 2008. "Le Cassiquiare, . . .Rio entre Rios . . . Un Fleuve entre Deux Fleuves." *Bulletin de Liaison des Membres de la Société de Géographie* 1(1529):9–39.

Giacottino, J.-C. 1984. *Les Guyanes*. Collection Que sais-je? Paris: Presses Universitaires de France.

Glaser, B. 2007. "Prehistorically Modified Soils of Central Amazonia a Model for Sustainable Agriculture in the Twenty-first Century." *Philosophical Transactions of the Royal Society of London, Series B, Biological Sciences* 362:187–196.

Glaser, B., L. Haumaier, G. Guggenberger, and W. Zech. 2001. "The 'Terra Preta' Phenomenon: A Model for Sustainable Agriculture in the Humid Tropics." *Naturwissenschaften* 88:37–41.

Glaser, B., and W. I. Woods (eds.). 2004. *Amazonian Dark Earths: Explorations in Space and Time*. Heidelberg: Springer.

Gondard, P., and F. López. 1983. *Inventario Arqueológico Preliminar de los Andes Septentrionales del Ecuador*. Quito: MAG/PRONAREG/ORSTOM, Banco Central del Ecuador.

Goodland, E. A. 1964. "The Mound." *Journal of the British Guiana Museum and Zoo* 39:9–17.

Goulding, M., R. Barthem, and E. Ferreira. 2003. *The Smithsonian Atlas of the Amazon*. Washington, D. C., and London: Smithsonian Books.

Goupy-des-Marets, J. 1690. *Voyage aux Iles d'Amérique et aux Côtes d'Afrique en 1675 et 1676*. Bibliothèque Municipale de Rouen (Ms Montbret 125).

Grenand, F. 1996. "'Abattis contre essart, *again*'." *Journal d'Agriculture Traditionnelle et de Botanique Appliquée* 38(1):19–53.

Grenand, P. 1971. *Les Relations Intertribales en Haute Guyane du XVII*[ème] *Siècle à nos Jours*. Paris: Maîtrise Spécialisée d'Ethnologie.

———. 1979. "Commentaires à propos d'un Abattis Wayapi (Guyane Française)." *Cahiers ORSTOM, Série Sciences Humaines*, XVI(4):299–304.

———. 1980. *Introduction à l'Étude de l'Univers Wayãpi: Ethnoécologie des Indiens du Haut-Oyapock (Guyane française)*. Langues et Civilisations à Tradition Orale No. 40. Paris: SELAF/CNRS/ORSTOM.

———. 1981. "Agriculture sur Brûlis et Changements Culturels: le Cas des Indiens Wayãpi et Palikur de Guyane." *Journal d'Agriculture Traditionnelle et de Botanique Appliquée* 28(1):23–31.

———. 1982. *Ainsi Parlaient nos Ancêtres: Essai d'Ethnohistoire Wayapi*. Paris: ORSTOM.

Grenand, P., and F. Grenand. 1987. "La Côte d'Amapá, de la Bouche de l'Amazonie à la Baie d'Oyapock, à Travers la Tradition Orale Palikur." *Boletim do Museu Paraense Emilio Goeldi, Nova Série, Antropologia*, 3(1):1–77.

———. 1993. "Histoire du Peuplement de la Várzea en Amazonas." *Amazoniana* 12(3/4):506–526.

Grenand, P., C. Moretti, H. Jacquemin, and M.-F. Prévost. 2004. *Pharmacopées Traditionnelles en Guyane (Créoles, Palikur, Wayãpi)*. Paris: IRD.

Groupe de Recherche et d'Échanges Technologiques [GRET]. 1990. *Manuel d'Agronomie Tropicale Appliquée à l'Agriculture Haïtienne*. Paris: Groupe de Recherche et d'Échanges Technologiques/Faculté d'Agronomie et de Médecine Vétérinaire d'Haïti.

Grimaldi, M. 1990. "Pédologie des Barres Prélittorales de la Plaine Côtière Ancienne." *Symposium International PICG 274/ORSTOM, Guide de l'Excursion B: les Plaines de Kourou et de Sinnamary, Pédologie* 37–44.

Grimaldi, M., B. Barthès, and C. Grimaldi. 1992. "Variations Spatiales du Sol des Barres Prélittorales de la Plaine Côtière de Guyane." In *Évolution des Littoraux de Guyane et de la Zone Caraïbe Méridionale Pendant le Quaternaire*, pp. 251–264. Collection Colloques et Séminaires. Paris: Éditions de l'ORSTOM.

Guapindaia, V. 2001. "Encountering the Ancestros: The Maracá Urns." In *Unknown Amazon*, edited by C. McEwan, C. Barreto, and E. Neves, pp. 156–173. London: The British Museum Press.

Guisan, S. 1788. *Traité sur les Terres Noyées de la Guiane Appelées Communément Terres Basses, sur leur Dessèchement, leur Culture et l'Exploitation de leurs Productions, avec des Réflexions sur la Régie des Esclaves et Autres Objets*. Paris: Imprimerie du Roy.

Gumilla, J. (1745) 1963. *El Orinoco Ilustrado y Defendido*. Caracas: Biblioteca de la Academia Nacional de la Historia.

Hammond, D. S. 2005. *Tropical Forests of the Guiana Shield: Ancient Forests in a Modern World*. Oxford/Cambridge: Cabi Publishing.

Hammond, D. S., and H. ter Steege. 1998. "Propensity of Fire in Guianan Rainforests." *Conservation Biology* 12:944–947.

Hammond, D. S., H. ter Steege, and K. van der Borg. 2006. "Upland Soil Charcoal in the Wet Tropical Forests of Central Guyana." *Biotropica* 39:153–160.

Hammond, N. 1986. "Le Développement de la Civilisation Maya." *Pour la Science* 108:16–25.

Hansen, É., and A. Le Dreff. 2002. "Oiseaux d'Eau du Littoral Guyanais." In *Guyane ou le Voyage Écologique*, edited by C. Richard-Hansen and R. Le Guen, pp. 202–217. Garies: Éditions Roger Le Guen.

Harcourt, R. (1613) 1926. *A Voiage of Discoverie to Guiana Performed by Robert Harcourt in 1609*. Second Series, LX. London: The Hakluyt Society.

Harrison, P. D., and B. L. Turner II (eds.). 1978. *Pre-hispanic Maya Agriculture*. Albuquerque: University of New Mexico Press.

Hartt, C. F. 1871a. "Brazilian Rock Inscriptions." *American Naturalist* 5(3):139–157.

———. 1871b. "The Ancient Indian Pottery of Marajo, Brazil." *American Naturalist* 5(5):259–271.

Heide, J. van der. 1973. *De Hertenrits: Een Bodemkunding/Archeologisch*. Rapport Dienst Bodemkartering No. 43. Paramaribo.

Heckenberger, M. J. 2005. *The Ecology of Power: Culture, Place and Personhood in the Southern Amazon, AD 1000–2000*. New York: Routledge

———. 2009. "Lost Cities of the Amazon." *Scientific American* 301:64–71.

Heckenberger, M. J., A. Kuikuro, U. T. Kuikuro, J. C. Russell, M. Schmidt, C. Fausto, and B. Franchetto. 2003. "Amazonia 1492: Pristine Forest or Cultural Parkland?" *Science* 301(5640):1710–1714.

Heckenberger, M. J., J. C. Russell, C. Fausto, J. R. Toney, M. J. Schmidt, E. Pereira, B. Franchetto, and A. Kuikuro. 2008. "Pre-Columbian Urbanism, Anthropogenic Landscapes, and the Future of the Amazon." *Science* 321:1214–1217.

Hemming, J. 1978. *The Search for El Dorado*. London: Phoenix Books.

———. 1987. *Amazon Frontier: The Defeat of the Brazilian Indians*. London: Pan Macmillan.

———. 1990. *Roraima: Brazil's Northernmost Frontier*. Institute of Latin American Studies Research Papers. London: University of London.

———. 2003. *Die If You Must: The Brazilian Indians in the Twentieth Century*. London: Pan Macmillan.

———. 2008. *Tree of Rivers: The Story of the Amazon*. London: Thames and Hudson.

Hilbert, P. P. 1968. *Archäologische Untersuchungen am Mittleren Amazonas*. Marburger Studien zur Völkerkunde 1. Berlin: Dietrich Reimer.

Hiraoka, M. 1986. "Zonation of Mestizos Riverine Farming Systems in Northeast Peru." *National Geographic Research* 2:354–371.

Hiraoka, M., and S. Mora (eds.). 2001. *Desarrollo Sostenible en la Amazonía ¿Mito o realidad?*. Hombre y Ambiente 63–64. Quito: Abya-Yala.

Hoock, J. 1971. *Les Savanes Guyanaises: Kourou. Essai de Phytoécologie Numérique*. Mémoires ORSTOM No. 44. Paris: ORSTOM.

Howard, G. D. 1947. "Prehistoric Ceramic Styles of Lowland South America, Their Distribution and History." *Yale University Publications in Anthropology* 37:5–95.

Huber, W. 1887. "Rapport de la Médaille d'Or à M. Jean Chaffanjon." *Bulletin de la Société de Géographie* 409–418.

Hurault, J.-M. 1965. *La Vie Matérielle des Noirs Réfugiés Boni et Indiens Wayana du Haut-Maroni (Guyane Française): Agriculture, Économie et Habitat*. Paris: ORSTOM.

———. 1989. *Français et Indiens en Guyane (1604–1972)*. Cayenne: Guyane Presse Diffusion.

———. 2000. "Montagnes Mythiques: Les Tumuc-Humac." *Cahiers de l'Outre-Mer* 53(212):367–392.

Im Thurn, E. F. (1883) 1967. *Among the Indians of Guiana*. New York: Dover Publications.

Imazio, Silveira da. 1994. *Estudo Sobre Estrategias de Subsistencia de Cacadores-coletores Pre-historicos do Sitio Gruta du Gavião, Caruja (Pará)*. [Master's thesis], Arqueologia, Universidade de São Paulo.

Institute Nationale de Statistics and Economic Studies [INSEE]. 1962. *Résultats Statistiques du Recensement Général de la Population des Départements d'Outre-Mer Effectué le 9 Octobre 1961: Guyane*. Paris: Imprimerie Nationale.

Iriarte, J. 2007. "New Perspectives on Plant Domestication and the Development of Agriculture in the New World." In *Rethinking Agriculture: Archaeological and Ethnoarchaeological Perspectives*, edited by T. Denham, J. Iriarte, and L. Vrydaghs, pp. 167–187. Walnut Creek, California: Left Coast Press.

Iriarte, J., B. Glaser, J. Watling, A. Wainwright, J. J. Birk, S. Rostain, and D. McKey. 2010. "Late Holocene Neotropical Agricultural Landscapes: Phytolith and Stable Carbon Isotope Analysis of Raised Fields from French Guianan Coastal Savannas." *Journal of Archaeological Science* 37:2984–2994.

Iriarte, J., M. Power, S. Rostain, F. Mayle, H. Jones, J. Watling, B. Whitney, and D. McKey. 2012. "Fire-free Land Use in Pre-1492 Amazonian Savannas Amazonia." *Proceedings of the National Academy of Sciences* 109(17):6473–6478.

Jackson, J. 1983. *The Fish People: Linguistic Exogamy and Tukanoan Identity in the Northwest Amazon*. Cambridge: Cambridge University Press.

Janin, T. 2002. "Awala-Yalimapo: Étude Anthropologique de l'Urne Funéraire." *Bilan Scientifique de la Région Guyane 1999, No. 41*. Cayenne: DRAC, Service Régional de l'Archéologie.

Jolivet, M.-J. 1982. *La Question Créole, Essai de Sociologie sur la Guyane Française*. Collection Mémoire 96. Paris: Éditions de l'ORSTOM.

———. 2002. "Images de Guyane, entre Réduction et Cloisonnement." *Autrepart* 24:107–124.

Kelly, E. F., S. W. Blecker, C. M. Yonker, C. G. Olson, E. E. Wohl, and L. C. Todd. 1998. "Stable Isotope Composition of Soil Organic Matter and Phytoliths as Paleoenvironmental Indicators." *Geoderma* 82:59–81.

Kendall, A., and A. Rodríguez. 2009. *Desarollo y Perspectivas de los Sistemas de Andenerias en los Andes Centrales del Peru*. Lima: CBC/IFEA.

Kern, D. 1988. *Caracterizasão Pedólogica de Solos com Terra Preta Arqueológica na Região de Oriximiná, Para* [M.A. thesis]. Universidade Federal do Rio Grande do Sul, Porto Alegre.

Keymis, L. (1596) 1722, "Relation de la Guiane Traduite de l'Anglois du Capitaine Keymis (en 1596)." In *Voyage de François Corréal aux Indes Occidentales*, pp. 261–288. Amsterdam.

Khubadux, M. R., G. J. R. Maat, and A. H. Versteeg. 1991. "The Remains of Prehistoric Amerindians of the Tingi Holo Ridge in Suriname: A Physical Anthropological Investigation of the Versteeg Collection." *Comptes Rendus du 12e Congrès de l'Association Internationale d'Archéologie des Caraïbes*, Martinique, pp. 135–151.

Klinken, G. J. van. 1991. *Dating and Dietary Reconstruction by Isotopic Analysis of Amino Acids in Fossil Bone Collagen, with Special Reference to the Caribbean*. Publications Foundation for Scientific Research in the Caribbean Region 128, Amsterdam.

Knapp, G. 1991. *Andean Ecology: Adaptative Dynamics in Ecuador*. Dellplain Latin American Studies, 27. Boulder: Westview Press.

Knapp, G., and R. Ryder. 1983. "Aspects of the Origin, Morphology and Function of Ridged Fields in the Quito Altiplan, Ecuador." In *Drained Field Agriculture in Central and South America*, edited by J. P. Darch, pp. 201–220. BAR International Series 189. Oxford: Archeopress.

Kolata, A. L. 1991. "The Technology and Organization of Agricultural Production in the Tiwanaku State." *Latin American Antiquity* 2:99–125.

———. 1996. *Tiwanaku and Its Hinterland: Archaeology and Paleoecology of an Andean Civilization, 1: Agroecology*. Washington, D. C.: Smithsonian Intitution Press.

Kolata, A. L., O. Rivera, J. C. Ramirez, and E. Gemio. 1996. "Rehabilitating Raised-field Agriculture in the Southern Lake Titicaca Basin of Bolivia." In *Tiwanaku and Its Hinterland: Archaeology and Paleoecology of an Andean Civilization, 1: Agroceology*, edited by A. L. Kolata, pp. 203–230. Washington, D. C.: Smithsonian Intitution Press.

Krook, L. 1979. *Sediment Petrographical Studies in Northern Suriname*. Amsterdam: Elinkwijk.

La Condamine, C.-M. de. (1745) 1981. *Voyage sur l'Amazone*. Collection La Découverte. Paris: Maspéro.

Lathrap, D. W. 1970. *The Upper Amazon*. London: Thames and Hudson.

———. 1973. "The Antiquity and Importance of Long Distance Trade Relations in the Moist Tropics of pre-Columbian South America." *World Archaeology* 5:170–186.

Lathrap, D. W., and J. Oliver. 1987. "Agüerito: el Complejo Poícromo más Antiguo de Américaen la Confluencia del Apure y Orinoco." *Interciencia* 12(6):274–289.

Lavallée, D. 2005. "Néolithisations en Amérique: Des Prédateurs Semi-nomades aux Sociétés Complexes." *Annales Histoires, Sciences Sociales* 5:1035–1067.

Leblond, J.-B. 1813. *Description Abrégée de la Guyane Française*. Paris.

Lehmann, J., J. Gaunt, and M. Rondon. 2006. "Bio-char Sequestration in Terrestrial Ecosystems: A Review." *Mitigation and Adaptation Strategies for Global Change* 11:403–427.

Lehmann, J., D. C. Kern, B. Glaser, and W. I. Woods (eds.). 2003. *Amazonian Dark Earths: Origin, Properties, Management*. Dordrecht: Kluwer Academic Publishers.

Leite, E.-C., R. Kasisi, and P. Jacobs. 2003. "Stratégies de Gestion Durable des Écosystèmes Aquatiques en Afrique: le Cas du Complexe du Lac Nokoué-lagune de Porto Novo au Bénin." In *Euro-Mediterranean Information System on the Know-how in Water Sector*, pp. 1–9. Kyoto: EMWIS.

Lelong, M. 1953. *Le Fleuve des Carajas*. Paris: Julliard.

Lemaire, A. 1981. "Le Pays d'Eden et le Bit-Adini aux Origines d'un Mythe." *Syria* 58(3–4):313–330.

Léotard, G., A. Duputie, F. Kjellberg, E. J. P. Douzery, C. Debain, J.-J. de Granville, and D. McKey. 2009. "Phylogeography and the Origin of Cassava: New Insights from the Northern Rim of the Amazonian Basin." *Molecular Phylogenetics and Evolution* 53:329–334.

Le Roux, J. 2006. *La Guyane, le Bagne: Un Siècle de Presse Illustrée de 1840 à 1940*. Condé-sur-Noireau: Éditions du Valhermeil.

Le Roux, Y. 1992. "La Révolution Agricole des Terres Basses en Guyane." In *Évolution des Littoraux de Guyane et de la Zone Caraïbe Méridionale pendant le Quaternaire*, pp. 327–346. Collection Colloques et Séminaires. Paris: Éditions de l'ORSTOM.

Leroyer, C., J.-E. Brochier, and S. Thiébault. 2008. "L'Environnement Végétal: Néolithisation et Transformations du Paysage Végétal: Rythmes et Modalités." In

Archéologie de la France: Le Neolithique, edited by J. Tarête and C.-T. Le Roux, pp. 53–76. Paris: Éditions Picard/Ministère de la Culture et de la Communication.

Lescallier, D. 1799. *Tableau de Cayenne ou de la Guïane Française*. Paris: Tilliard & Fils Libraires.

Létard, R. 2007. *L'Enfant de la Mangrove*. Guyane: Auto-Edition.

Lévi-Strauss, C. 1948. "The Tribes of the Upper Xingu River." In *Handbook of South American Indians, Volume 4: The Circum-Caribbean Tribes*, edited by J. Steward, pp. 321–348. Bureau of American Ethnology Bulletin 143. Washington, D. C.: Smithsonian Institution.

———. 1994. *Saudades do Brasil*. Paris: Librairie Plon.

Lézy, E. 1989. *Guyane, de l'Autre Côté des Images*. Paris: L'Harmattan.

———. 2000. *Guyane, Guyanes: Une géographie "Sauvage" de l'Orénoque à l'Amazone*. Collection Mappemonde. Paris: Belin.

———. 2003. "Introduction." In *Cahiers de l'Amérique Latine* 43:13–18.

Lima, H., E. Góes Neves, and J. B. Petersen. 2006. "La Fase Açutuba: um Novo Complexo Cerâmico na Amazônia Central." *Arqueología Suramericana* 2(1):26–52.

Lombardo, U., E. Canal-Beeby, S. Fehr, and H. Veit. 2011. "Raised Fields in the Bolivian Amazonia: A Prehistoric Green Revolution or a Flood Risk Mitigation Strategy?" *Journal of Archaeological Science* 38(3):502–512.

Lombardo, U., and H. Prümers. 2010. "Pre-Columbian Human Occupation Patterns in the Eastern Plains of the Llanos de Mojos, Bolivian Amazonia." *Journal of Archaeological Science* 37(8):1875–1885.

Loy, T. H. 1994. "Methods in the Analysis of Starch Residues on Prehistoric Stone Tools." In *Tropical Archaeobotany: Applications and New Developments*, edited by J. G. Hather, pp. 86–114. London: Routledge.

Lozet, J., and C. Mathieu. 1986. *Dictionnaire de Science du Sol: Technique et Documentation*. Paris: Lavoisier.

Madella, M., M. K. Jones, P. Goldberg, Y. Goren, and E. Hovers. 2002. "The Exploitation of Plant Resources by Neanderthals in Amud Cave (Israel): The Evidence from Phytolith Studies." *Journal of Archaeological Science* 29:703–719.

Mam Lam Fouck, S. 1999. *La Guyane Française au Temps de l'Esclavage, de l'Or et de la Francisation (1802–1946)*. Collection Espaces Guyanais. Cayenne: Ibis Rouge Éditions.

Mann, C. C. 2005. *1491: New Revelations of the Americas before Columbus*. New York: A. A. Knopf.

Mayle, F. E., R. P. Langstroth, R. A. Fisher, and P. Meir. 2007. "Long-term Forest-savanna Dynamics in the Bolivian Amazon: Implications for Conservation." *Phil Trans Royal Society London* 362(1478):291–307.

Mazière, M., and G. Mazière. 1993. "Maripasoula, Montagne Couronnée de Yaou." In *Bilan scientifique de la région Guyane, 1992*, is there any (eds)?? pp. 21–23. Cayenne: DRAC Guyane, Service Régional de l'Archéologie, Ministère de la Culture et de la Francophonie.

McClain, M. E., and R. E. Cossío. 2003. "The Use of Riparian Environments in the Rural Peruvian Amazon." *Environmental Conservation* 30(3):242–248.

McEwan, C., C. Barreto, and E. Neves (eds.). 2001. *Unknown Amazon*. London: The British Museum Press.

McKey D., and S. Beckerman. 1993. "Chemical Ecology, Plant Evolution, and the Evolution of Traditional Manioc Cultivation Systems." In *Tropical Forests, People and Food: Biocultural Interactions and Applications to Development*, edited by C. M. Hladik, A. Hladik, O. F. Linares, H. Pagezy, A. Semple, and M. Hadley, pp. 83–112. Paris: UNESCO, Parthenon.

McKey, D., S. Rostain, J. Iriarte, B. Glaser, J. Birk, I. Holst, and D. Renard. 2010. "Pre-Columbian Agricultural Landscapes, Ecosystem Engineers and Self-organized Patchiness in Amazonia." *Proceedings of the National Academy of Sciences* 107(17):7823–7828.

Meggers, B. J. 1966. *Ecuador*. Ancient Peoples and Places No. 49. London: Thames and Hudson.

———. 1971. *Amazonia, Man and Culture in a Counterfeit Paradise*. Studies in Cultural Ecology. Los Angeles: University of California Press.

———. 1994. "Archaeological Evidence for the Impact of Mega-Niño Events on Amazonia during the Past Two Millennia." *Climatic Change* 28:321–338.

———. 1996. *Amazonia, Man and Culture in a Counterfeit Paradise* (revised edition). Washington, D. C./London: Smithsonian Institution Press.

———. 2003. "Natural versus Anthropogenic Sources of Amazonian Biodiversity: The Continuing Quest for El Dorado." In *How Landscapes Change*, edited by G. Bradshaw and P. Marquet, pp. 89–107. Berlin: Springer Verlag.

———. 2007. "Mid-Holocene Climate and Cultural Dynamics in Brazil and the Guianas." In *Climatic Change and Cultural Dynamics: A Global Perspective on mid-Holocene Transitions*, edited by D. G. Anderson, K. A. Maasch, and D. H. Sandweiss, pp. 117–155. Amsterdam: Elsevier.

Meggers, B. J., and C. Evans. 1957. *Archaeological Investigations at the Mouth of the Amazon*. Bureau of American Ethnology Bulletin 167. Washington, D. C.: Smithsonian Institution.

———. 1961. "An Experimental Formulation of Horizon Styles in the Tropical Forest Area of South America." In *Essays in Pre-columbian Art and Archaeology* edited by S. K. Lothrop et al., pp. 372–388. Cambridge: Harvard University Press.

Meggers, B. J., C. Evans, and E. Estrada. 1965. *Valdivia and Machalilla Phases*. Washington, D. C.: Smithsonian Institution.

Mercader, J., F. Runge, L. Vrydaghs, H. Doutrelepont, C. E. N. Ewango, and J. Juan-Tresseras. 2000. "Phytoliths from Archaeological Sites in the Tropical Forest of Ituri, Democratic Republic of Congo." *Quaternary Research* 54:102–112.

Mestre, M., and S. Delpech. 2008. *Plateau des Mines (Saint-Laurent du Maroni), Rapport de Fouille*. Cayenne: INRAP.

Métraux, A. 1959. "La Révolution de la Hache." *Diogène* 25:32–45.

Millet Cámara, L. 1984. "Logwood and Archaeology in Campeche." *Journal of Anthropological Research* 40(2):324–328.

Mollard, É., and A. Walter. 2008. *Agricultures Singulières*. Paris: IRD Éditions.

Montabo, B. 2004. *Le Grand Livre de l'Histoire de la Guyane*. Sainte-Clotilde: Orphie.

Montoroi, J.-P. 1996. *Gestion Durable des Sols de la Mangrove au Sénégal en Période de Sécheresse: Dynamique de l'Eau et Géochimie des Sels d'un Bassin Versant Aménagé*. Collection Études et Thèses. Paris: Éditions de l'ORSTOM.

Moore, J. 1988. "Prehistoric Raised Field Agriculture in the Casma Valley, Peru." *Journal of Field Archaeology* 15:265–276.

Mora, S., L. F. Herrera, I. Cavelier, and C. Rodríguez. 1991. *Cultivars, Anthropic Soils and Stability: A Preliminary Report of Archaeological Research in Araracuara, Colombian Amazon*. Latin American Archaeology Reports 2. Pittsburgh: University of Pittsburgh.

Morey, N. K. 1975. *Ethnohistory of the Colombian and Venezuelan Llanos* [Ph.D. dissertation]. Department of Anthropology, University of Utah. Available at University of Michigan Dissertation Information Service, Ann Arbor, Michigan.

Morris, A. 2004. *Raised Field Technology: The Raised Fields Project Around Lake Titicaca*. King's SOAS Studies in Development Geography. Ashgate: Burlington.

Mothes, P. (ed.). 1998. *Actividad Volcánica y Pueblos Precolombinos en el Ecuador*. Quito: Ediciones Abya-Yala.

Muse, M., and F. Quintero. 1987. "Experimentos de Reactivación de Campos Elevados, Peñón del Río, Guayas, Ecuador." In *Pre-Hispanic Agricultural Fields in the Andean Region*, Part 1, edited by W. Denevan, D. Mathewson, and G. Knapp, pp. 249–266. BAR International Series 359(i). Oxford: Archeopress.

Natural History of South America: Guiana Shield, Great American Interchange, Gondwana, Nazca Plate, South American Plate, Peru-Chile Trench. 2010. Memphis: Books LLC.

Neves, E. G. 2006. *Arqueologia da Amazônia*. Collection Descobrindo o Brasil. Rio de Janeiro: Jorge Zahar.

Neves, E. G., and J. B. Petersen. 2006. "The Political Economy of Pre-Columbian Amerindians: Landscape Transformation in Central Amazonia." In *Time and Complexity in Historical Ecology*, edited by W. Balée and C. L. Erickson, pp. 279–309. New York: Columbia University Press.

Neumann, K. 2003. "New Guinea: A Cradle of Agriculture." *Science* 301(5630):180-181.

Niederberger Betton, C. 1987. *Paléopaysages et Archéologie Pré-urbaine du Bassin du Mexique*. Mexico City: CEMCA.

Nimuendaju, C. (1926) 2008. *Les Indiens Palikur et Leurs Voisins*. Encyclopédie Palikur 1. Orléans: CTHS.

Nordenskiöld, E. (1916) 2009. "Indian Adaptations in Flooded Regions of South America." *Journal of Latin American Geography* 8(2):209–224.

———. 1929. *Analyse Ethno-géographique de la Culture Matérielle de Deux Tribus Indiennes du Gran Chaco*. Études d'Ethnographie Comparée I. Paris: Éditions Genet.

———. 1930. *L'Archéologie du Bassin de l'Amazone*. Ars Americana 1. Paris.

National Planning Office of Suriname and Regional Development and Physical Planning Department [NPOS and RDPPD]. 1988. *Suriname Planatlas*. Washington, D. C.: NPOS and RDPPD.

Nunn, Patrick D. 2000. "Environmental Catastrophe in the Pacific Islands around AD 1300." *Geoarchaeology: An International Journal* 15(7):715–740.

Oliver, J. 2001. "The Archaeology of Forest Foraging and Agricultural Production in Amazonia." In *Unknown Amazon*, edited by C. McEwan, C. Barreto, and E. Neves, pp. 50–85. London: The British Museum Press.

———. 2008. "The Archaeology of Agriculture in Ancient Amazonia." In *Handbook of South American Archaeology*, edited by H. Silverman and W. Isbell, pp. 185–216. New York: Springer/Kluwer/Plenum.

Olsen, K. M., and B. A. Schaal. 1999. "Evidence on the Origin of Cassava: Phylogeography of *Manihot esculenta*." *Proceedings of the National Academy of Sciences*. 96:5586–5591.

Oyuela-Caycedo, A., and R. M. Bonzani. 2005. *San Jacinto 1: A Historical Ecological Approach to an Archaic Site in Colombia*. Tuscaloosa: University of Alabama Press.

Padoch, C., J. M. Ayres, M. Pinedo-Vasquez, and A. Henderson (eds.). 1999. *Várzea: Diversity, Development, and Conservation of Amazonia's Whitewater Floodplains*. Advances in Economic Botany 13. New York: New York Botanical Garden.

Pagan, F. J.-B. de. 1655. *Relation Historique et Géographique de la Grande Rivière des Amazones Extraite de Divers Auteurs et Réduite en Meilleure Forme avec la Carte d'Icelle Rivière et de ses Provinces*. Paris: C. Besongue.

Palmatary, H. C. 1960. "The Archaeology of the Lower Tapajós Valley, Brazil." *Transactions of the American Philosophical Society* 50(3).

Parsons, J. J. 1969. "Ridged Fields in the Río Guayas Valley, Ecuador." *American Antiquity* 34:76–80.

Parsons, J. J., and W. A. Bowen. 1966. "Ancient Ridged Fields in the San Jorge River Floodplain." *Geographical Review* 56:317–343.

Parsons, J. J., and W. M. Denevan. 1967. "Pre-Columbian Ridged Fields." *Scientific American* 217(1):92–100.

Pärssinen, M., and A. Korpisaari. 2003. *Western Amazônia*. Renval Institute Publications 14. Helsinki: University of Helsinki.

Pärssinen, M., D. Schaan, and A. Ranzi. 2009. "Pre-Columbian Geometric Earthworks on the Upper Purús: A Complex Society in Western Amazonia." *Antiquity* 83(322):1084–1095.

Pauw, C. de. 1770. *Recherches Philosophiques sur les Américains, ou Mémoires Intéressants pour Servir à l'Histoire de l'Espèce Humaine*. Berlin.

Péan, C. 1948. *Conquête en Terre de Bagne*. Paris: Éditions Altis.

Pearsall, D. M., and D. R. Piperno (eds.). 1993. *Current Research in Phytolith Analysis: Applications in Archaeology and Paleoecology*. MASCA Research Papers in Science and Archaeology 10. Philadelphia: University Museum of Archaeology and Anthropology, University of Pennsylvania.

Pélissier, P. 1966. *Les Paysans du Sénégal: Les Civilisations Agraires du Cayor à la Casamance*. Saint-Yrieix: Imprimerie Fabrègue.

Peltier, W. R., and R. G. Fairbanks. 2006. "Global Glacial Ice Volume and Last Glacial Maximum Duration from an Extended Barbados Sea Level Record." *Quaternary Science Reviews* 25:3322–3337.

Pessenda, L. C. R., S. E. M. Gouveia, R. Aravena, R. Boulet, and E. P. E. Valencia. 2004. "Holocene Fire and Vegetation Changes in Southeastern Brazil as Deduced from Fossil Charcoal and Soil Carbon Isotopes." *Quaternary International* 114:35–43.

Pessenda, L. C. R., S. E. M. Gouveia, R. Aravena, B. M. Gomes, R. Boulet, and S. Ribeiro. 1998. "^{14}C Dating and Stable Carbon Isotopes of Soil Organic Matter in Forest-savanna Boundary Areas in the Southern Brazilian Amazon Region." *Radiocarbon* 40:1013–1022.

Petersen, J. B., E. Neves, and M. J. Heckenberger. 2001. "Gift From the Past: *Terra Preta* and Prehistoric Amerindian Occupation in Amazonia." In *Unknown Amazon*, edited by C. McEwan, C. Barreto, and E. Neves, pp. 86–105. London: The British Museum Press.

Petitjean Roget, H. 1991. "50 Sites de Montagnes en Guyane Française: Contribution à l'Inventaire Archéologique d'Emile Abonnenc." In *Comptes Rendus du XIIème Congrès de l'Association Internationale d'Archéologie des Caraïbes*, Martinique, pp. 241–258.

Pétrequin, P., and A.-M. Pétrequin. 1993. *Écologie d'un Outil: la Hache de Pierre en Irian Jaya (Indonésie)*. Monographie du CRA 12. Paris: CNRS Éditions.

Petrullo, V. 1939. *Archaeology of Arauquin*. Bureau of American Ethnology Bulletin 123, pp. 291–295. Washington, D. C.: Smithsonian Institution.

Petry Cabral, M., and J. Darcy da Moura Saldanha. 2009. "Note sur des Structures Mégalithiques en Guyane Brésilienne." *Journal de la Société des Américanistes* 95(1):97–110.

Pierre, M. 1982. *La Terre de la Grande Punition: Histoire des Bagnes de Guyane*. Paris: Ramsay.

Piperno, D. R. 1991. "The Status of Phytolith Analysis in the American Tropics." *Journal of World Prehistory* 5:155–191.

———. 1995. "Plant Microfossils and Their Application in the New World Tropics." In *Archaeology in the Lowland American Tropics: Current Analytic Methods and Recent Applications*, edited by P. W. Stahl, pp. 130–153. Cambridge: Cambridge University Press.

Piperno, D. R., and T. D. Dillehay. 2008. "Starch Grains on Human Teeth Reveal Early Broad Crop Diet in Northern Peru," *Proceedings of the National Academy of Sciences* 105:19622–19627.

Piperno, D. R., and I. Holst. 1998. "The Presence of Starch Grains on Prehistoric Stone Tools from the Humid Neotropics: Indications of Early Tuber Use and Agriculture in Panama." *Journal of Archaeological Science* 25:765–776.

Piperno, D. R., and D. M. Pearsall. 1998. "The Silica Bodies of Tropical American Grasses: Morphology, Taxonomy and Implications for Grass Systematics and Fossil Phytolith Identification." Smithsonian Contributions to Botany 85. Washington, D.C.: Smithsonian Institution Press.

Piperno, D. R., A. J. Ranere, I. Holst, and P. Hansell. 2000. "Starch Grains Reveal Early Root Crop Horticulture in the Panamanian Tropical Forest." *Nature* 407:894–897.

Pires, J. M., and G. T. Prance. 1985. "The Vegetation Types of the Brazilian Amazon." In *Key Environments: Amazonia*, edited by G. T. Prance and T. E. Lovejoy, pp. 109–145. Oxford: Pergamon Press.

Pitou, L.-A. 1805. *Voyage Forcé à Cayenne, dans les Deux Amériques et Chez les Anthropophages*. Paris.

Plazas, C., and A. M. Falchetti de Saenz. 1981. *Asentamientos Prehispanicos en el Bajo Río San Jorge*. Fundación de Investigaciones Arqueológicas Nacionales. Bogotá: Banco de la República.

———. 1987. "Poblamiento y Adecuación Hidráulica en el Bajo Río San Jorge, Costa Atlantica, Colombia." In *Pre-Hispanic Agricultural Fields in the Andean Region*, edited by W. Denevan, K. Mathewson, and G. Knapp, pp. 483–503. BAR International Series No. 359(ii). Oxford: Archeopress.

Plazas, C., A. M. Falchetti de Saenz, J. Saenz Samper, and S. Archilla. 1993. *La Sociedad Hidráulica Zenú: Estudio Arqueológico de 2000 Años de Historia en las Llanuras del Caribe Colombiano*. Bogotá: Banco de la República.

Plew, M. G. 2005. *The Archaeology of Guyana*. Bar International Series No. 1400. Oxford: Archeopress.

Politis, G. G. 2001. "Foragers of the Amazon: The Last Survivors or the First to Succeed?" In *Unknown Amazon*, edited by C. McEwan, C. Barreto, and E. Neves, pp. 26–49. London: The British Museum Press.

———. 2007. *Nukak: Ethnoarchaeology of an Amazonian People*. Walnut Creek: Left Coast Press.

Porras, P. 1987. *Investigaciones Arqueológicas a las Faldas del Sangay*. Centro de Investigaciones Arqueológicas. Quito: PUCE.

Porro, A. 1985. "Mercadorias e Rotas de Commercio Intertribal na Amazônia." *Revista do Museu Paulista* 30:7–12.

Posey, D. A. 1985. "Indigenous Management of Tropical Forest Ecosystems: The Case of the Kayapó Indians of the Brazilian Amazon." *Agroforestry Systems* 3:139–158.

Pozorski, T., and S. Pozorski. 1982. "Reassessing the Chicama-Moche Intervalley Canal: Comments on Hydraulic Engineering Aspects of the Chimu Chicam-Moche Intervalley Canal." *American Antiquity* 47:851–868.

Préfontaine, B. de. 1763. *Maison Rustique a l'Usage Des Habitans de la France Équinoxiale, Connue sous le Nom de Cayenne*. Paris: C. J. B. Bauche Libraire.

Prost, G. 1967. "Dans le Nord-Est du Brésil." *Les Cahiers d'Outre-Mer* 21:78–102.

Prost, M.-T. 1987a. *Projet de Recherches sur l'Environnement Côtiers des Guyanes*. ORSTOM/Pôle Halieutique et Océanologique Caraïbe. Cayenne: Centre ORSTOM de Cayenne.

———. 1987b. "Shoreline Changes in French Guiana." *Quaternary of South America and Antarctic Peninsula* 5:291–299.

———. 1990a. "Beaches and Cheniers in French Guiana." *Quaternary of South America and Antartic Peninsula* 6:189–219.

———. 1990b. *Les Côtes des Guyanes*. Programme Environnement Côtier, Série Géomorphologie-Sédimentologie. Cayenne: Centre ORSTOM de Cayenne.

———. 1990c. "Modifications des Rivages et Conditions Actuelles de Formation des Cheniers sur les Côtes de Cayenne." In *Symposium International PICG 274/ ORSTOM, Guide de l'excursion A: Presqu'île de Cayenne*, pp. 16–30. Cayenne: Géomorphologie.

———. 1992. "Sédimentation Côtière et Formation de Cheniers en Guyane: la Zone de Cayenne." In *Évolution des Littoraux de Guyane et de la Zone Caraïbe Méridionale Pendant le Quaternaire*, pp. 397–414. Collection Colloques et Séminaires. Paris: Éditions de l'ORSTOM.

Prümers, H. 2004. "Hügel Umgeben von 'Schönen Monstern': Ausgrabungen in der Loma Mendoza (Bolivien)." *Expeditionen in Vergessene Welten. 25 Jahre archäologische Forschungen in Amerika, Afrika und Asien* 10:47–78.

Puleston, D. E. 1977. "The Art and Archaeology of Hydraulic Agriculture in the Maya Lowlands." In *Social Procès in Maya Prehistory: Studies in Memory of Sir Eric Thompson*, edited by N. Hammond, pp. 449–467. New York: Academic Press.

Puleston, D. E. 1978. "Terracing, Raised Fields, and Tree Cropping in the Maya Lowlands: A New Perspective on the Geography of Power." In *Pre-hispanic Maya Agriculture*, edited by P. Harrison and B. Turner II, pp. 225–245. Albuquerque: University of New Mexico Press.

Raleigh, W. (1596) 1993. *El Dorado*. Collection UNESCO d'Oeuvres Représentatives. Paris: Éditions Utz.

Ranzi, A. 2008. *Paleontologia da Amazônia: Mamíferos Fósseis do Juruá*. Rio Branco: M. M. Paim.

Redmond, E. M., and C. S. Spencer. 1998. "Prehispanic Causeways and Regional Politics in the Llanos of Barinas, Venezuela." *Latin American Antiquity* 9(2):95–110.

———. 2007. *Archaeological Survey in the High Llanos and Andean Piedmont of Barinas, Venezuela*. Anthropological Papers of the American Museum of Natural History 86. New York: American Museum of Natural History.

Reichel-Dolmatoff, G. 1985. *Monsú, un Sitio Arqueológico*. Bogotá: Biblioteca Banco Popular, Textos Universitarios.

Reichel-Dolmatoff, G., and A. Reichel-Dolmatoff. 1974. "Un Sistema de Agricultura Prehistórica de los Llanos Orientales." *Revista Colombiana de Antropología* 17:189–200.

Renard, D. 2007. *Anciens Remaniements des Paysages et Fonctionnement des Écosystèmes: Géométrie des Paysages, Structure de la Communauté Végétale et Écologie du Mutualisme de la Dispersion* Manihot tristis/Ectatomma brunneum *dans une Savane Saisonnièrement Inondée de Guyane Française* [M.A. thesis]. Montpellier: Université de Montpellier-II.

Renard, D. 2010. Histoire et Écologie des Complexes de Champs Surélevés dans les Savanes Côtières de Guyane Française [PhD dissertation]. Université de Montpellier-2, Montpellier.

Renard, D., J. Iriarte, J. J. Birk, S. Rostain, B. Glaser, and D. McKey. 2011. "Ecological Engineers Ahead of Their Time: The Functioning of Pre-Columbian Raised-field Agriculture and Its Potential Contributions to Sustainability Today." *Ecological Engineering* doi:10.1016/J.ecoleng.2011.03.007.

Renard, D., J. J. Birk, B. Glaser, J. Iriarte, G. Grisard, J. Karl, and D. McKey. 2012. "Origin of Mound-field Landscapes: A Multi-proxy Approach Combining Contemporary Vegetation, Carbon Stable Isotopes and Phytoliths." *Plant and Soil* doi:10.1007/s11104-011-0967-8.

Renoux, F., M. Fleury, Y. Reinette, P. Grenand, and F. Grenand. 2003. "L'Agriculture Itinérante sur Brûlis dans les Bassins du Maroni et de l'Oyapock: Dynamique et Adaptation aux Contraintes Spatiales." *Revue Forestière Française*, "Connaissance et Gestion de la Forêt Guyanaise," edited by M. Fournier and J. Weigel, 55:236–259.

Ribeiro, D. 2002. *Carnets Indiens: Avec les Indiens Urubus-Kaapor, Brésil.* Collection Terre Humaine. Paris: Plon.

Rietkerk, M., S. C. Dekker, P. C. de Ruiter, and J. van de Koppel. 2004. "Self-organized Patchiness and Catastrophic Shifts in Ecosystems." *Science* 305:1926–1929.

Rietkerk M., and J. van de Koppel. 2008. "Regular Pattern Formation in Real Ecosystems." *Trends in Ecology and Evolution* 23(3):169–175.

Rivière, P. 1984. *Individual and Society in Guiana: A Comparative Study of Amerindian Social Organization.* Cambridge Studies in Social Anthropology. Cambridge: Cambridge University Press.

Roeleveld, W., and A. J. van Loon. 1979. "The Holocene Development of the Young Coastal Plain." *Geologie en Mijnbouw* 58:21–28.

Roosevelt, A. C. 1980. *Parmana, Prehistoric Maize and Manioc Subsistence along the Amazon and Orinoco.* New York: Academic Press.

———. 1991. *Moundbuilders of the Amazon: Geophysical Archaeology on Marajó Island, Brazil.* New York: Academic Press.

———. 1995. "Early Pottery in the Amazon." In *The Emergence of Pottery: Technology and Innovation in Ancient Societies*, edited by W. K. Barnett and J. W. Hoopes, pp. 115–131. Washington, D. C./London: Smithsonian Institution Press.

———. 1997. *The Excavations at Corozal, Venezuela: A Stratigraphy and Ceramic Seriation.* Yale University Publications in Anthropology 83. New Haven: Yale University Press.

Roosevelt, A. C., R. A. Housley, M. Imazio da Silveira, S. Maranca, and R. Johnson. 1991. "Eighth Millenium Pottery from a Prehistoric Shell Midden in the Brazilian Amazon." *Science* 254:1621–1624.

Roosevelt, A. C., M. Lima da Costa, C. Lopes Machado, M. Michab, N. Mercier, H. Valladas, J. Feathers, W. Barnett, M. Imazio da Silveira, A. Henderson, J. Sliva, B. Chernoff, D. S. Reese, J. A. Holman, N. Toth, and K. Schick. 1996. "Paloindian Cave Dwellers in the Amazon: The Peopling of the Americas." *Science* 272:373–384.

Rosique T., F. Pous, and P. Charles-Dominique. 2000. "Évolution Morphogénique Holocène d'un Bassin Versant de la Forêt Guyanaise: la Nourague Occidentale (Guyane Française)." *Comptes Rendus de l'Académie des Sciences* Series IIA "Earth and Planetary Science" 330(5):333–340.

Rossetti, D. de F., A. M. Góes, and P. Mann de Toledo. 2009. "Archaeological Mounds in Marajó Island in Northern Brazil: A Geological Perspective Integrating Remote Sensing and Sedimentology." *Geoarchaeology: An International Journal* 24(1):22–41.

Rostain, S. 1990. "Étude d'une Chaîne Opératoire: Les Haches en Pierre Polie d'Amazonie." *Arquivos do Museu de História Natural* 11:195–240.

———. 1991. *Les Champs Surélevés Amérindiens de la Guyane.* Collection La Nature et l'Homme. Cayenne: ORSTOM.

———. 1992. "La Céramique Amérindienne de Guyane Française." *Bulletin de la Société Suisse des Américanistes* 55–56:93–127.

—. 1994a. *L'Occupation Amérindienne Ancienne du Littoral de Guyane.* TDM 129. Paris: ORSTOM Éditions.

———. 1994b. "The French Guiana Coast: A Key Area in Prehistory between the Orinoco and Amazon Rivers." *Between St. Eustatius and the Guianas* 3:53–97.

———. 1994c. "Archéologie du Littoral de Guyane, une Région Charnière entre les Influences Culturelles de l'Orénoque et de l'Amazone." *Journal de la Société des Américanistes* 80:9–46.

———. 1995. "La Mise en Culture des Marécages Littoraux de Guyane à la Période Précolombienne Récente." In *Milieux, Sociétés et Archéologues*, edited by A. Marliac, pp. 119–160. Paris: ORSTOM/Khartala.

———. 1996. "La Peau des Défunts." *Totem* 15:4.

———. 1997. *Arqueología del Río Upano, Amazonía Ecuatoriana.* Quito: IFEA.

———. 1999a. "Occupations Humaines et Fonction Domestique de Monticules Préhistoriques en Haute Amazonie Équatorienne." *Bulletin de la Société Suisse des Américanistes* 63:71–95.

———. 1999b. "Secuencia Arqueológica en Montículos del Valle del Upano en la Amazonía Ecuatoriana." *Bulletin de l'Institut Français d'Études Andines* 28(1):1–37.

———. 2003. "Cinq Petits Tapirs: Les Guyanes Amérindiennes d'avant 1499." *Cahiers de l'Amérique Latine* 43:19–37.

———. 2006a. "Etnoarqueología de la Casa Huapula y Jívaro." *Bulletin de l'Institut Français d'Études Andines*, Número Temático "Avances de Investigación en el Ecuador Prehispánico," edited by M. Guinea and J.-F. Bouchard. 35(3):337–346.

———. 2006b. "Spécialisation et Commerce dans les Guyanes Amérindiennes." *Techniques et Cultures* 46–47:139–174.

———. 2007. "Histoire de la Recherche Archéologique dans les Guyanes Côtières." *Les Nouvelles de l'Archéologie*, dossier "L'archéologie des Départements Français d'Amérique," edited by S. Rostain and N. Vidal. 108–109:36–56.

———. 2008a. "Le littoral des Guyanes, Héritage de l'Agriculture Précolombienne." *Études Rurales* 181:9–38.

———. 2008b. "Agricultural Earthworks on the French Guiana Coast." In *Handbook of South American Archaeology*, edited by H. Silverman and W. Isbell, pp. 217–234. New York: Springer/Kluwer/Plenum.

———. 2008c. "The Archaeology of the Guianas: An Overview" In *Handbook of South American Archaeology*, edited by H. Silverman and W. Isbell, pp. 279–302. New York: Springer/Kluwer/Plenum.

———. 2009a. "Between Orinoco and Amazon: The Ceramic Age In The Guianas." In *Anthropologies of Guayana: Cultural Spaces in Northeastern Amazonia*, edited by N. Whitehead and S. Alemán, pp. 36–54. Tucson: Arizona University Press.

Rostain, S. 2009b. "Western Coast of the Guianas around 1000 BP." *Proccedings of the 22th Congress of the International Association for Caribbean Archaaeology*, Kingston.

———. 2010a. "Pre-Columbian Earthworks in Coastal Amazonia." *Diversity*, Special Issue "Long-Term Anthropic Influences on the Diversity of Amazonian Landscapes and Biota," edited by W. Balée, 2(3):353–369. Available at www.mdpi.com/journal/diversity.

———. 2010b. "Archéologie de l'Ouest Guyanais: Le Site de Sable Blanc." In *Amaz'hommes: Sciences de l'Homme et Sciences de la Nature en Amazonie*, edited by E. Barone-Visigalli and A. Roosevelt, pp. 59–81. Collection Espace Outre-Mer. Matoury: Ibis Rouge Éditions.

———. 2010c. "Cacicazgos Guyanenses: Mito o Realidad?" In *Proceedings of the 1ero Encontro Internacional de Arqueologia Amazônica*, pp. 168–192. Belém: MPEG.

———. 2012. "Between Sierra and Selva: Pre-Columbian Landscapes in the Upper Ecuadorian Amazonia." *Quaternary International*, Special Issue "Human Occupation of Tropical Rainforests," edited by Norm Catto, 249:31–42.

Rostain, S., L. Déodat, and L. Clerc. 2009. *Bois Diable, Rapport de la Campagne de Fouilles Programmées 2008*. Nanterre: CNRS/Université de Panthéon-Sorbonne.

Rostain, S., N. Guillaume-Gentil, and L. Clerc. 2008. *Sable Blanc Est, Rapport de Fouille Programmée*. Nanterre: CNRS, UMR 8096.

Rostain, S., and A. H. Versteeg. 2003. "Recherche sur l'Archéologie de la Côte Occidentale de Guyane." *Journal de la Société des Américanistes* 89(1):161–175.

———. 2004. "The Arauquinoid Tradition in the Guianas." In *Late Ceramic Societies in the Eastern Caribbean*, edited by A. Delpuech and C. Hofman, pp. 233–250. BAR 1273, Monographs in American Archaeology 14, Paris.

Rostain, S., and Y. Wack. 1987. "Haches et Herminettes en Pierre de Guyane Française." *Journal de la Société des Américanistes* 73:107–138.

Roth, W. E. 1924. *An Introductory Study of the Arts, Crafts and Customs of the Guiana Indians*. 38th Annual Report of Bureau of American Ethnology. Washington, D. C.: Smithsonian Institution.

Rouse, I., and J. M. Cruxent. 1963. *Venezuelan Archaeology*. New Haven/London: Yale University Press.

Saavedra Arteaga, O. 2006. "El Sistema Agrícola Prehispánico de Camellones en la Amazonía Boliviana." In *Agricultura Ancestral: Camellones y Albarradas*, edited by F. Valdez, pp. 295–311. Quito: Ambassade de France, IFEA, IRD, CNRS, INPC, Banco Central del Ecuador, Abya-Yala, Université de Paris-1.

Salazar, E. 2008. "Pre-Columbian Mound Complexes in the Upano River Valley, Lowland Ecuador." In *Handbook of South American Archaeology*, edited by H. Silverman and W. Isbell, pp. 263–278. New York: Springer/Kluwer/Plenum.

Sanaiotti, T. M., L. A. Martinelli, R. L. Victoria, S. E. Trumbore, and P. B. Camargo. 2002. "Past Vegetation Changes in Amazon Savannas Determined Using Carbon Isotopes of Soil Organic Matter." *Biotropica* 34:2–16.

Sanders, W. T., J. R. Parsons, and R. Santley. 1979. *The Basin of Mexico: Ecological Processes in the Evolution of Civilization*. New York: Academic Press.

Sanoja Obediente, M. 1979. *Las Culturas Formativas del Oriente de Venezuela: La Tradición Barrancas del Bajo Orinoco*. Serie "Estudios, monografías y ensayos" 6. Caracas: Biblioteca de la Academia Nacional de la Historía.

Saunaluoma, S., and D. Schaan. 2012. "Monumentality in Western Amazonian Formative Societies: Geometric Ditched Enclosures in the Brazilian State of Acre." *Antiqua* 1:e1. doi:10.4081/antiqua.2012.e1.

Scarborough, V. L., and L. J. Lucero. 2010. "The Non-hierarchical Development of Complexity in the Semitropics: Water and Cooperation." *Water History* Special Issue "Ancient Near East and Americas" 2(2):185–205.

Schaan, D. P. 1997. *A Linguagem Iconográfica da Cerâmica Marajoara: um Estudo da Arte Pré-histórica na Ilha de Marajó, Brasil (400–1300 AD)*. Coleção Arqueologia 3. Porto Alegre: EDIPUCRS.

———. 2000. "Recent Investigations on Marajoara Culture, Marajó Island, Northern Brazil." *Antiquity* 74:469–470.

———. 2004. *The Camutins Chiefdom: Rise and Development of Complex Societies on Marajó Island, Brazilian Amazon* [Ph.D. dissertation]. Department of Anthropology, University of Pittsburgh, Pennsylvania.

———. 2008. "The Nonagricultural Chiefdoms of Marajó Island." In *Handbook of South American Archaeology*, edited by H. Silverman and W. Isbell, pp. 339–357. New York: Springer.

———. 2012. *Sacred Geographies of Ancient Amazonia: Historical Ecology of Social Complexity*. Walnut Creek: Left Coast Press.

Schaan, D., A. Ranzi, and M. Pärssinen. 2008. *Arqueologia do Acre: do Pronapaba às Pesquisas Sobre os Geoglifos. Arqueologia da Amazônia Ocidental. Os Geoglifos do Acre*. Belém: EDUFPA.

Schnier, M. 2009. *Raised Fields in French Guiana: Between Science and Local Knowledge* [B.A. thesis]. Department of Social Anthropology, University of Bayreuth, Germany.

Schoepf, D. 1972. "Historique et Situation Actuelle des Indiens Wayana-Aparai du Brésil." *Bulletin du Musée d'Ethnographie de Genève* 15:33–64.

Seurin, M. 1976. "Étude d'un Cordon Littoral de l'Anse de Rémire (Guyane)." *Connaissance de la Guyane* 3:6–22.

Silman, M. R. 2007. "Plant Species in Amazonian Forests." In *Tropical Rainforest Response to Climatic Change*, edited by M. B. Bush and J. R. Flenley, pp. 267–294. New York: Springer.

Simões, M. F. 1981. "Coletores-pescadores Ceramistas do Litoral do Salgado (Pará)." *Boletim do Museu Paraense Emilio Goeldi Série Antropologia* 78:1–32.

Simon, G. n.d. *Field Notes and Map Regarding Raised Fields Features near Fort Nassau*. Georgetown: Walter Roth Museum of Anthropology.

Soubiès, F. 1979–80. "Existence d'une Phase Sèche en Amazonie Brésilienne Datée par la Présence de Charbons dans les Sols (6.000-3.000 ans BP)." *Cahiers de l'ORSTOM, Série Géologie* 9:133–146.

Souty, F. J.-L. 1982. "Agriculture et Système Agricole au Suriname de la Fin du XVIIe à la Fin du XVIIIe siècle." *Revue Française d'Histoire d'Outre-mer* 69(256):193–223.

Spencer, C. S., E. M. Redmond, and M. Rinaldi. 1994. "Drained Fields at La Tigra, Venezuelan Llanos: A Regional Perspective." *Latin American Antiquity* 5(2):95–110.

Stedman, J. G. (1796) 1992. *Stedman's Surinam: Life in an Eighteenth-century Slave Society. An Abridged, Modernized Edition of Narrative of a Five Year Expedition Against the Revolted Negroes of Surinam*, edited by R. Price and S. Price. Baltimore: The Johns Hopkins University Press.

Stemper, D. M. 1993. *The Persistence of Prehispanic Chiefdoms on the Río Daule, Coastal Ecuador*. University of Pittsburgh Memoirs in Latin American Archaeology 7. Pittsburgh/Quito: Ediciones Libri Mundi.

Steward, J. H. (ed.). 1948. *Handbook of South American Indians, Volume 3: The Tropical Forest Tribes*. Bureau of American Ethnology Bulletin 143. Washington, D. C.: Smithsonian Institution.

Strathern, Andrew. 1971. *The Rope of Moka: Big Men and Ceremonial Exchange in Mount Hagen, New Guinea.* Cambridge: Cambridge University Press.

Swartley, L. 2002. *Inventing Indigenous Knowledge: Archaeology, Rural Development and the Raised Field Rehabilitation Project in Bolivia.* Indigenous Peoples and Politics Series, edited by F. Wilmer. NewYork/London: Routledge.

Tacoma, J., D. C. Geijskes, G. J. R. Maat, and G. N. van Vark. 1991. *On "Amazonidi" Precolumbian Skeletal Remains and Associated Archaeology from Suriname.* Publications of the Foundation for Scientific Research in the Caribbean Region 127, Amsterdam.

Tardy, C. 1998. *Paléoincendies Naturels, Feux Anthropiques et Environnements Forestiers de Guyane Française du Tradiglaciaire à l'Holocène Récent: Approche Chronologique et Anthracologique* [Ph.D. dissertation]. Department of Biology of Populations and Ecology Université Montpellier-II, France.

Tardy, C., M. Mestre, S. Jérémie, and J. Briand. 2010. "Réflexions sur la Dynamique à Long Terme des Écosystèmes Amazoniens: Paléoincendies Naturels, Feux Anthropiques et Occupations Humaines: Recherches Paleoenvironnementales en Guyane Française." In *Amaz'hommes: Sciences de l'Homme et Sciences de la Nature en Amazonie*, edited by E. Barone-Visigalli and A. Roosevelt, pp. 105–108. Collection Espace Outre-Mer. Matoury: Ibis Rouge Éditions.

Thomas, J. 2002. "Awala-Yalimapo: Village et Abords." *Bilan Scientifique de la Région Guyane 1998* Cayenne: DRAC.

Thoms, A. V., 2003. "Cook-stone Technology in North America: Evolutionary Changes in Domestic Fire Structures during the Holocene." *Le Feu Domestique et ses Structures au Néolithique et aux Âges des Métaux*, edited by M.-C. Frère-Sautot, Collection Préhistoires, 9:87–96.

———. 2008. "The Fire Stones Carry: Ethnographic Records and Archaeological Expectations for Hot-rock Cookery in Western North America." *Journal of Anthropological Archaeology* 27(4):443–460.

Tony, C. 1842. "Voyage dans l'Intérieur de la Guyane chez les Indiens Roucoyens (1769)." *Nouvelles Annales des Voyages* pp. 213–235.

Torre, C. de la., and M. Burga. 1986. *Andenes y Camellones en el Peru Andino: Historia Presente y Futuro.* Lima: CONYTEC.

Torres, C. M. 1987. *The Iconography of South American Snuff Trays and Related Paraphernalia.* Etnologiska Sudier 37. Göteborg: Göteborgs Etnografiska Museum.

Transler, A.-L., P. Saffache, and D. Moullet. 2006. "Impacts des Activités Agricoles sur le Littoral Guyanais." *La Géographie*, spécial "Antilles-Guyane," hors-série 1522:70–77.

Turenne, J.-F. 1979. "Sédimentologie des Plaines Côtières." In *Atlas des Départements d'Outre-Mer no. IV: la Guyane.* Bordeaux-Talence: CEGET-CNRS/ORSTOM.

———. 2002. *D'Argile et de Terres.* Fort-de-France: Exposition IRD.

———. 2008. "La Vision Exotique de la Fertilité, Contribution à l'Histoire de la Guyane." *Mondes et Cultures: Compte Rendu Annuel des Travaux de l'Académie des Sciences d'Outre-mer* 68(1):163– 177.

Turner, B. L., and P. D. Harrison. 1981. "Prehistoric Raised-field Agriculture in the Maya Lowlands." *Science* 213(4506):399–405.

Uhl, C., and J. Sadarriaga. 1987. "La Fragilité de la Forêt Amazonienne." *Pour la Science* 119:38–47.

Vacher S., S. Jérémie, and J. Briand (eds.). 1998. *Amérindiens du Sinnamary (Guyane): Archéologie en Forêt Équatoriale.* Paris: Maison des Sciences de l'Homme, Daf 70.

Valdez, F. 2008a. "La Laguna de la Ciudad, le grenier de La Tolita." *Les Nouvelles de l'archéologie*, special "Des mers de glace à la terre de feu: L'archéologie française en Amérique," edited by S. Rostain. Paris: Éditions de la Maison des Sciences de l'Homme/éditions Errance. 111–112:69–75.

———. 2008b. "Inter-zonal Relationships in Ecuador." In *Handbook of South American Archaeology*, edited by H. Silverman and W. Isbell, pp. 865–888. New York: Springer/Kluwer/Plenum.

Van den Bel, M. 2006. *Les Occupations Amérindiennes du Site Eva 2 (Sinnamary), Rapport Final de Fouille*. Cayenne: INRAP.

———. 2009. "Les Dépôts Funéraires d'Iracoubo: Les Résultats d'une Fouille Préventive d'une Nécropole Précolombienne en Guyane Française." *Amazônica, Revista de Antropologia* 1(1):230–249.

Van der Hammen, T., and M. L. Absy. 1994. "Amazonia during the Last Glacial." *Palaeogeography, Palaeoclimatology, Palaeoecology* 109:247–261.

Van der Hammen, T., and H. Hooghiemstra. 1996. "The History of the Forests and Climate of Tropical South America." In *The Americas: The Conservation Atlas of Tropical Forests*, pp. 17–26. New York: Simon & Schuster.

Vargas Arena, I. 1979. *La Tradición Saladoide del Oriente de Venezuela: La Fase Cuartel*. Serie "Estudios, Monografías y Ensayos" 5. Caracas: Biblioteca de la Academia Nacional de la Historía.

———. 1981. *Investigaciones Arqueológicas en Parmana: Los Sitios de La Gruta y Ronquín, Estado Guarico, Venezuela*. Serie "Estudios, Monografías y Ensayos" 20. Caracas: Biblioteca de la Academia Nacional de la Historía.

Veblen, T. T., K. R. Young, and A. R. Orme. 2007. *The Physical Geography of South America*. Oxford: Oxford University Press.

Vernet, J.-L. 1980. "La Végétation du Bassin de l'Aude, entre Pyrénées et Massif Central, au Tardiglaciaire et au Postglaciaire d'après l'Analyse Anthracologique." *Review of Palaeobotany and Palynology* 30:33–55.

Versteeg, A. H. 1978. "A Distinctive Kind of Pottery in Western Suriname." *Mededelingen Surinaams Museum* 23–24:16–27.

———. 1981. "A Fortified Pre-columbian Village in East Suriname?" *Mededelingen Surinaams Museum* 32:39–56.

———. 1983. "Raised Field Complexes and Associated Settlements in the Coastal Plain of Western Suriname." In *Drained Field Agriculture in Central and South America*, edited by J. P. Darch, pp. 237–250. BAR International Series 189. Oxford: Archeopress.

———. 1985. "The Prehistory of the Young Coastal Plain of West Suriname." *Ber Rijksdienst Oudheidkundig Bodemonderzoek* 35:653–750.

———. 2003. *Suriname voor Columbus/Suriname before Columbus*. Libri Musei Surinamensis 1. Paramaribo: Stichting Surinaams Museum.

———. 2007. *Werehpai Archaeological Research*. Paramaribo: Stichting Surinaams Museum/Conservation International. Available at home.wxs.nl/~vrstg/guianas/werehpai.html.

Versteeg, A. H., and F. C. Bubberman. 1992. *Suriname before Columbus*. Paramaribo: Stichting Surinaams Museum, Mededelingen, 49a.

Versteeg, A. H., and S. Rostain (eds.). 1997. *The Archaeology of Aruba: the Tanki Flip Site*. Amsterdam/Aruba: Publications of the Archaeological Museum Aruba 8/Verre Naasten Naderbij.

Versteeg, A. H., J. Tacoma, and P. van de Velde. 1990. *Archaeological Investigations on Aruba: The Malmok Cemetery*. Amsterdam/Aruba: Publication of the

Archaeological Museum Aruba 2, publication of the Foundation for Scientific Research in the Caribbean Region 126.

Vilhena Vialou, A. 2005. *Pré-história do Mato Grosso, Vol. 1: Santa Elina.* São Paulo: EDUSP.

Villalba, M., and A. Alvarado. 1998. "La Arqueología del Valle de Quito en Clave Volcánica." In *Actividad Volcánica y Pueblos Precolombinos en el Ecuador*, edited by P. Mothes, pp. 73–110. Quito: Ediciones Abya-Yala.

Villalba Sevilla, J. F. F. 2007. *Estudio Funcional de los Camellones de Cayambe, Sistema Agrícola Precolombino, en el Sector La Tola, de la Sierra Norte del Ecuador (Período de Integración)* [tesis de licenciatura en arqueología]. Guayaquil: ESPOL.

Walker, J. H. 2004. *Agricultural Change in the Bolivian Amazon.* University of Pittsburgh Memoirs in Latin American Archaeology 13. Pittsburgh/Trinidad: Fundación Kenneth Lee.

Wassén, H. 1934. "The Frog-motive among the South American Indians. 1ère partie." *Anthropos* 29(3/4):319–370.

———. 1965. *The Use of Some Specific American Indian Snuff and Related Paraphernalia.* Göteborg: Etnologiska Sudier, Göteborgs Etnografiska Museum.

Webb, S. D. 1997. "The Great American Faunal Interchange." In *Central America: A Natural and Cultural History*, edited by A. G. Coates, pp. 97–122. New Haven: Yale University Press.

Weil, F. 2005. *Histoire de New York.* Paris: Fayard.

Werkhoven, M. C. M., and A. H. Versteeg. 1980. "The Vegetation of Four Mounds in the Coastal Plain of Suriname." *Mededelingen Surinaams Museum* 32:8–37.

Whitehead, N. 1993. "Historical Discontinuity and Ethnic Transformation in Native Amazonia and Guayana, 1500–1900." *L'Homme* 28:289–309.

———. 1994. "The Ancient Amerindian Polities of the Lower Orinoco, Amazon and Guayana Coast: A Preliminary Analysis of Their Passage from Antiquity to Extinction." In *Amazonian Indians: From Prehistory to the Present*, edited by A. Roosevelt, pp. 33–54. Tucson: University of Arizona Press.

———. 1996. "Searching for Paradise? Recent Research in Amazonian Archaeology." *Journal of Archaeological Research* 4(3):241–264.

———. 2002. *Dark Shamans: Kanaimà and the Poetics of Violent Death.* Durham/London: Duke University Press.

———. 2003. "The Sign of Kanaimá, the Space of Guayana and the Demonology of Development." *Cahiers de l'Amérique Latine*, Dossier "La Guyane, une Île en Amazonie." 43:67–86.

Whitehead, N., M. J. Heckenberger, and G. Simon. 2010. "Materializing the Past among the Lokono (Arawak) of the Berbice River, Guyana." *Antropológica* 54(114):87–127.

Whitmore, T. M., and B. L. Turner II. 2001. *Cultivated Landscapes of Middle America on the Eve of Conquest.* New York: Oxford University Press.

Williams, D. 1992. "El Arcaico en el Noroeste de Guyana y los Cominezos de la Horticultura." In *Prehistoria Sudamericana: Nuevas Perspectivas*, edited by B. J. Meggers, pp. 233–251. Washington, D. C.: Taraxacum.

———. 2003. *Prehistoric Guiana.* Kingston/Miami: Ian Randle Publishers.

Willis, K. J., L. Gillson, and T. M. Brncic. 2004. "How 'Virgin' Is Virgin Rainforest?" *Science* 304:402–403.

Wilson, C., I. A. Simpson, and E. J. Currie. 2002. "Soil Management in Pre-Hispanic Raised Field Systems: Micromorphological Evidence from Hacienda Zuleta, Ecuador." *Geoarchaeology: An International Journal* 17(3):261–283.

Wittmann, F., W. J. Junk, and M. T. F. Piedade. 2004. "The Várzea Forests in Amazonia: Flooding and the Highly Dynamic Geomorphology Interact with Natural Forest Succession." *Forest Ecology and Management* 196(2–3):199–212.

Woods, W. I. 1995. "Comments on the Black Earths of Amazonia." *Papers and Proceedings of Applied Geography Conferences* 18:159–165.

Woods, W. I., W. G. Teixeira, J. Lehmann, C. Steiner, A. M. G. A. WinklerPrins, and L. Rebellato (eds.). 2009. *Amazonia Dark Earths: Wim Soebroek's Vision*. New York: Springer.

Yde, J. 1965. *Material Culture of the Waiwai*. Ethnografisk Raekke X. Copenhagen: Nationalmuseets Skrifter.

Zola Baez, M. G. 1986. "Aménagement Préhispanique des Zones Inondées du Veracruz (Mexique)." *Cahiers des Sciences Humaines* 22(1):83–95.

Zucchi, A. 1975. *Caño Canori: Un Grupo Prehispanico de la Selva de los Llanos de Barinas*. Colección Antropología 5. Caracas: Universidad Central de Venezuela/ Facultad de Ciencias Economicas y Sociales.

———. 1991. "La Serie Meillacoide y sus Relaciones con la Cuenca del Orinoco." In *Proceedings of the 11th Congress of the International Association for Caribbean Archaeology*, Fundación Arqueológica, Antropologica e Historica de Puerto Rico/Universidad de Puerto Rico/United States Forest Service, pp. 272–286.

Zucchi, A., and W. M. Denevan. 1979. *Campos Elevados e Historia Cultural Prehispánica en los Llanos Occidentales de Venezuela*. Caracas: Universidad Católica Andres Bello/Instituto de Investigaciones Históricas.

Zucchi, A., and K. Tarble. 1982. "Evolución y Antiguedad de la Alfarería con Esponjilla en Agüerito, un Yacimiento del Orinoco Medio." *Indiana*, 7:183–199.

———. 1984. "Nuevos Datos Sobre la Arqueología Tardía del Orinoco: La Serie Valloïde." *Acta Cientifica Venezolana* 35(5/6):434–445.

Zucchi, A., K. Tarble, and E. J. Vaz. 1984. "The Ceramic Sequence and New TL and C-14 Dates for the Agüerito Site of the Middle Orinoco, Venezuela." *Journal of Field Archaeology* 11:155–180.

Index

Italicized numbers indicate figures.

About the Author

Stéphen Rostain was born in Brittany, on the western coast of France. He is the first French archaeologist to dedicate his work to the Amazonian rainforest. After his first contacts with the New World archaeology in Teotihuacan, Mexico, and in the Maya site of Chaguite, Guatemala, he focused his research on Amazonian lowlands and Caribbean Islands. He was employed as an archaeologist in French Guiana from 1985 until he received his Ph.D. at the Pantheon-Sorbonne University in Paris in 1994. During this period he developed innovative aerial archaeology along the Guiana coast, where he discovered thousands of pre-Columbian raised fields. From 1994 to 1996, he was co-director of the archaeological project Tanki Flip in Aruba, which led to the publication of a large monograph. After the organization (with D. Schoepf) of the exhibition *Autoportraits du Nouveau Monde* in Geneva, Switzerland, he lived in Ecuador between 1996 and 2001, where he managed an archaeological project in the Upano Valley, particularly in the prestigious site of Sangay, famous for its numerous artificial earthmounds. His main interests are historical ecology of the Amazon, pre-Columbian earthworks, indigenous agriculture, and ethnoarchaeology. One original characteristic of his projects is an interdisciplinary approach with an extensive team of specialists.

Rostain has published more than 100 articles, book chapters, and books. He is author of *L'archéologie* (coedited with Y. Le Roux, 1990), *Les Champs Surélevés Amérindiens de la Guyane* (1991), *L'Occupation Amérindienne Ancienne du Littoral de Guyane* (1994), *The Archaeology of Aruba: the Tanki Flip Site* (coedited with A. H. Versteeg, 1997), *El Chagüite, Jalapa: El Período Formativo en el Oriente de Guatemala* (coedited with A. Ichon, R. Grignon, A. Blanco Padilla, and V. Gervais, 2000) and *Precolombiana* (coedited with A. H. Versteeg, 2005).

Rostain is presently Director of Research at the National Center for Scientific Research of France (CNRS), and for the last 10 years has been in charge (with D. McKey) of an interdisciplinary project on the Guiana coast. In 2008, he received the Paris Clio award for French projects in foreign countries. He is the director of the French Institute for Andean Studies (IFEA) in Ecuador and chairman of the 3rd International Congress of Amazonian Archaeology in Quito in 2013.